Local Energy Governance

T0256307

Local Energy Governance: Opportunities and Challenges for Renewable and Decentralised Energy in France and Japan examines the extent of the energy transition taking place at a local level in France and Japan, two countries that share ambitious targets regarding the reduction of GHG emissions, their share of renewable energy and their degree of market liberalization. This book observes local energy policies and initiatives and applies an institutional and legal analysis to help identify barriers but also opportunities in the development of renewable energies in the territories. The book will highlight governance features that incubate energy transition at the local level through interdisciplinary contributions that offer legal, political, sociological and technological perspectives. Overall, the book will draw conclusions that will also be informative for other countries aiming at promoting renewable energies. This book will be of great interest to students and scholars of energy policy and energy governance.

Magali Dreyfus is a Researcher at CNRS (French National Centre for Scientific Research). She is based at CERAPS (Center for European Research on Administration, Politics and Society) – Lille University in France. Formerly, she was a Visiting Fellow at GRIPS (National Graduate Institute for Policy Studies) and a Research Fellow at United Nations University Institute for the Advanced Study of Sustainability (UNU-IAS), in Tokyo, Japan.

Aki Suwa is a Professor at the Faculty of Contemporary Society, Kyoto Women's University (KWU), and an Instructor at the College of Policy Science, Ritsumeikan University. Her main responsibility at these institutions is to conduct research on environment and sustainable policy and approach, based on community and regional analysis in Japan and in Asia. Prior to her appointment she was a Research Fellow at the United Nations University Institute of Advanced Studies (UNU-IAS). Her research focuses on how environmental understanding facilitate policy formulation, and how this subsequently influences the degree of the policy implementation on ground. Prior to working in UNU-IAS, she completed a doctorate from the University College London, the University of London. She also holds a MSc degree in the Environmental Technology and Policy from the Imperial College of Science, Technology and Medicine.

Routledge Explorations in Energy Studies

Sustainable Energy Education in the Arctic
The Role of Higher Education
Gisele M. Arruda

Electricity and Energy Transition in Nigeria
Norbert Edomah

Renewable Energy Uptake in Urban Latin America
Sustainable Technology in Mexico and Brazil
Alexandra Mallett

Energy Cooperation in South Asia
Utilizing Natural Resources for Peace and Sustainable Development
Mirza Sadaqat Huda

Perspectives on Energy Poverty in Post-Communist Europe
Edited by George Jiglau, Anca Sinea, Ute Dubois and Philipp Biermann

Dilemmas of Energy Transitions in the Global South
Balancing Urgency and Justice
Edited by Ankit Kumar, Johanna Höffken and Auke Pols

Assembling Petroleum Production and Climate Change in Ecuador and Norway
Elisabeth Marta Tómmerbakk

International Law and Renewable Energy Investment in the Global South
Avidan Kent

Local Energy Governance
Opportunities and Challenges for Renewable and Decentralised Energy in France and Japan
Edited by Magali Dreyfus and Aki Suwa

For more information about this series, please visit: www.routledge.com/ Routledge-Explorations-in-Energy-Studies/book-series/REENS

Local Energy Governance

Opportunities and Challenges for Renewable and Decentralised Energy in France and Japan

Edited by Magali Dreyfus and Aki Suwa

Routledge
Taylor & Francis Group

LONDON AND NEW YORK

from Routledge

First published 2022
by Routledge
2 Park Square, Milton Park, Abingdon, Oxon OX14 4RN

and by Routledge
605 Third Avenue, New York, NY 10158

Routledge is an imprint of the Taylor & Francis Group, an informa business

© 2022 selection and editorial matter, Magali Dreyfus and Aki Suwa; individual chapters, the contributors

British Library Cataloguing-in-Publication Data
A catalogue record for this book is available from the British Library

Library of Congress Cataloging-in-Publication Data
A catalog record has been requested for this book

ISBN: 978-0-367-45891-1 (hbk)
ISBN: 978-1-032-21453-5 (pbk)
ISBN: 978-1-003-02596-2 (ebk)

DOI: 10.4324/9781003025962

Typeset in Times New Roman
by Deanta Global Publishing Services, Chennai, India

Contents

List of figures   xi
List of tables   xiv
List of contributors   xv
Acknowledgments   xvii

Introduction   1
MAGALI DREYFUS AND AKI SUWA

Context   1
Defining local energy governance: Why local energy governance
    is important   2
Why France and Japan?   4
Questions and chapters in the book   6
Notes   12
References   12

PART I
National Framework of Energy Governance   15

1 Searching for alternatives to fossil- and fission-based energy
  sources in France   17
GUILLAUME DEZOBRY AND MAGALI DREYFUS

The European energy and climate institutional framework   17
An institutional framework designed for centralized governance
    and nuclear hegemony in the French energy mix   20
The promotion and slow rise of other energy sources and
    decentralized governance   25
The French energy mix   25
Conclusion   28
Notes   28
References   30

2 **Energy transition in Japan: The political landscape and civil society's contribution**   32

MASAYOSHI IYODA

Governmental preferences and their political background 33
Changes within the national government 34
A brief history of civil society's advocacy on climate and energy
* issues 35*
Civil society's actions to promote energy transition to
* renewables 37*
Conclusion 40
Notes 42
References 43

PART II
Local government powers in the energy sector   45

3 **Local authorities and energy in France: Increasing duties, limited means of action**   47

FRANÇOIS-MATHIEU POUPEAU

Local authorities and energy market regulation 47
Local authorities and sectoral policies 50
A local rise under control 56
Conclusion 59
Notes 60
References 61

4 **Local energy governance: The Japanese context, development and typology**   62

AKI SUWA

Connecting local energy governance with national wealth and
* benefits sharing discourses 64*
The strategic challenge of the Japanese energy transition 68
Local energy governance development: the emergence of new
* organizational archetypes in Japan 72*
Discussion and conclusion 78
References 82

5 **Barriers to renewable energy? A case analysis of the Garorim Bay tidal plant project in South Korea** 85
BOMI KIM AND YOUHYUN LEE

The emergence of the local autonomy system in South Korea 86
FIT to RPS: implementation of the RPS system 87
Environmental impact assessment 88
Case record 90
Discussion 97
References 99

PART III
Local partnerships for the development of renewable energy at the local level: citizens, communities and companies 101

6 **Local public companies, local authority shareholders and electricity: Rarely one, never two, always three** 103
MARIE-ANNE VANNEAUX

Local public electricity companies with majority public
shareholdings: the most common legal form 106
Minority public shareholding and local electricity policy, new
perspectives 110
Conclusion 114
Notes 117
References 120

7 **Analysis of the value added to local economies by municipal power suppliers in Japan** 122
KENJI INAGAKI AND TAKUO NAKAYAMA

Expectations on the community energy sector 122
Typology of municipal power suppliers 124
Analysis of value added to local economies by new municipal
power suppliers 132
Case studies: new municipal power suppliers subject to case
analysis 134
Discussion and conclusion 137
Acknowledgments 139
Notes 139
References 140

Part IV
Territories with 100% renewable energy

143

8 **The voluntary initiatives, "positive energy territory" and "positive energy territory for green growth", first steps toward decentralization of the French energy system?** 145
BLANCHE LORMETEAU

Innovative rules of governance are not easily absorbed by the central level   147
TEPOS: fruitful experimentation grounds for a hesitant state   150
The state's mistrust and the failure of the TEPCVs   152
The creation of ETCs: the end of a territorial approach to energy?   154
Conclusion   156
Notes   157
References   158

9 **The feasibility of a 100% renewable energy scenario at the village level in Japan from an economic standpoint** 161
TAKUO NAKAYAMA

Introduction: the potential for stimulating the local economy through a renewable energy business   161
The economic model to analyze local value added   162
The evaluation of the renewable energy business by the local value-added analysis model   163
The scenario to realize a 100% renewable energy local authority   172
The simulation of local value added for a 100% renewable energy local authority   175
Summary   176
References   178

10 **Actors, motives and social implications of 100% renewable energy territories in Austria and Germany** 181
LAURE DOBIGNY

Methods and case studies   182
Circumstances and actors of 100% RE territories   183
Motives of local RE transition   184
Social and economic implications   188
Conclusion   193
Note   194
References   194

Part V
**Technological issues in energy transition: market, grids and
smart cities**   197

11 **Digital and energy transition in French cities: limits and
    asymptote effects**   199
    RAPHAËL LANGUILLON-AUSSEL

    Introduction: energy, the neglected alpha *and* omega *of the
        French smart city?   199*
    *The actors of energy and digital tech, and their urban
        strategies in France   200*
    *Japanese smart communities and French smart cities:
        cross-cultural experiments through urban energy
        innovations   204*
    *Conclusion: for a multi-scalar governance and a life-cycle
        approach of French smart cities   209*
    Notes   210
    References   210

12 **Analysis of supply–demand balances in western Japan grids in
    2030: Integrating large-scale photovoltaic and wind energies:
    challenges in cross-regional interconnections**   212
    ASAMI TAKEHAMA AND MANABU UTAGAWA

    Introduction   212
    *Regulatory framework of renewable energies and grid integration:
        the national energy plan and historical trends of renewable
        energy development   212*
    Methods   225
    Simulation results   231
    Conclusions   241
    Notes   241
    References   243

13 **A glimpse into smart cities: opportunities for the development
    of energy cooperatives for citizens and businesses in Mexico**   245
    LUIS ROMÁN ARCINIEGA GIL

    Context of the Mexican energy industry   246
    *Mexico's state of play and commitments to climate change and
        clean energy by 2050   247*

*The challenge of inequality: tackling energy poverty through
    locally determined energy projects   248*
*Reconciling interests: the benefits of the cooperative model for
    industry and business   250*
Conclusions   251
References   253

Conclusion and avenues for further research   257
MAGALI DREYFUS AND AKI SUWA

A similar energy policy and market structure development   257
A slow and controlled opening of energy governance   257
Common multiple benefits as an incentive for local actors   258
Local companies, a key instrument for local actors   259
Avenues for further research   259
Summary   260

Index   263

Figures

1.1 Liberalization of the electricity market in (metropolitan) France   18
1.2 French electric generation mix in 2019   26
2.1 Media coverage including words of "100% renewable energy" (Searched in a news article database of Nikkei Telecon 21 by counting the amount of media coverage that included words which mean 100% renewable energy each year.)   41
3.1 Local authorities and the regulation of the energy sector   51
3.2 The jurisdictions of the French territorial collectivities in energy-climate   58
4.1 Local income deficit   67
4.2 Japanese electricity source transition   69
4.3 Japanese electricity market reform   70
4.4 Trends of power generation in Japan   71
4.5 Grid parity achievements   72
4.6 Trends of renewable energy capacity in Japan   73
4.7 Konan Ultra Power scheme   74
4.8 A business scheme in Waita, Kumamoto   76
4.9 Izumisano PPS electricity and financial transactions   79
4.10 Origin of electricity supply (Izumisano PPS)   79
4.11 Initial typology of local energy arrangement   81
6.1 Breakdown of majority public shareholding in SPCs   108
6.2 Involvement of the various types of private shareholders in the capital of 60 electricity SPCs as a percentage of the number of SPCs   109
6.3 Recurrence of combinations of local public shareholders in electricity SPCs - An analysis of 70 SPCs   116
7.1 Population trend in Miyama city (Source: Miyama city homepage)   134
7.2 Breakdown of the value added to the local economy by Miyama Smart Energy   135
7.3 Alternative proportions of local investment and local staff lead to a significant difference in value added to the local economy   136

7.4   Population trend in Hioki city (Source: Hioki city homepage)   136
7.5   Breakdown of the value added to the local economy by Hioki
       Energy   138
7.6   In-house performance of operations leads to a significant
       difference in value added to the local economy   138
9.1   The amount of local value added by a small-scale hydroelectric
       power plant, M (unit: million yen)   165
9.2   The amount of local value added by three small-scale
       hydroelectric power plants in the village (unit: million yen)   166
9.3   Attribution of local value added during the operation and
       maintenance (O&M) phase of the three power generation
       businesses in the village (after income tax)   167
9.4   The amount of local value added by the N Ohisama power
       generation plant (unit: million yen)   168
9.5   The total local value added of solar power generation business
       (unit: million yen)   169
9.6   Attribution of local value added during the operation phase of
       the solar power generation business (after employees' income)   169
9.7   The amount of local value added by the wood-fired boiler (340
       kW) (unit: million yen)   170
9.8   Changes in accumulative local value added by the wood-fired
       boiler (340 kW) (unit: million yen)   171
9.9   The total local value added by the three businesses in the
       village (unit: million yen). ※Accumulated value from the
       installation until the 20th year of operation. Source: Based on
       data from Nishiawakura village   172
9.10   Attribution of local value added by the three businesses in
       the village (after employees' income). ※The total local value
       added during the operation/maintenance phase: about 270
       million yen. Source: Based on data from Nishiawakura village   172
12.1   Share of electricity generation by source in IEA countries in
       2017. Source: IEA data   213
12.2   Long-term energy plan for electricity power source mix
       in Japan. Source: METI (Ministry of Economy Trade and
       Industry): Long-Term Energy Supply and Demand Outlook, 2015   214
12.3   Historical trends of electricity generation in Japan by source.
       Source: IEA data   215
12.4   Cumulative capacity of renewable energies in Japan before
       and after the FIT scheme. Source: Data obtained from METI.
       FIT data 2012–2020. Data before FIT is based on a METI
       document: the committee for large-scale installation of
       renewable energies and next generation power grid   216

12.5   The nine TSO control zones and geographical patterns of wind
        and solar energy potentials. Source: TSO zone map from the
        Organization for Cross-Regional Coordination of Transmission
        Operators (OCCTO), "Outlook of Electricity Supply-Demand
        and Cross-Regional Interconnection Lines" (2019). Solar
        and wind figures simplify geographical patterns of renewable
        potentials in each zone. Solar and wind figures are attached for
        easy understanding   220
12.6   The main structure of cross-regional interconnections between
        nine TSO grids. Source: Documents from the Ministry of
        Economy, Trade and Industry (METI): "The next generation
        grid network committee". Numbers in ( ) denote the maximum
        load demand in each TSO zone in winter 2019 and summer
        2020. Numbers next to interconnections denote line capacity
        (aggregated operational capacity) in August 2019   221
12.7   Operation modes and time schedule of daytime pump-up mode
        for pumped storage hydropower, EVs in charging mode and
        HPs in heating mode. Source: Takehama   231
12.8   Grid balance in the Kyushu zone, 1–7 May (high case). Source:
        Takehama   232
12.9   Grid balance in the Chugoku zone, 1–7 May (high case).
        Source: Takehama   232
12.10   Grid balance in the Shikoku zone, 1–7 May (high case).
        Source: Takehama   232
12.11   Grid balance in the Kansai-Chubu zone, 1–7 May (high case).
        Source: Takehama   233
12.12   Power transmission from the Kyushu, Shikoku and Chugoku
        zones to the Kansai-Chubu zone, 1–14 May. Source: Takehama   234
12.13   Maximum values of power transmission through inter-regional
        connections in May (high case). Source: Takehama   235
12.14   Flexible operations absorbing VRE power and power
        oversupply in Kyushu in May (high case). Source: Takehama   235
12.15   Control reserve activations in the Kyushu zone in the fourth
        week in May (high case). Source: Takehama   236
12.16   Grid balance in the Kyushu zone, 1–7 August (high case)   238
12.17   Grid balance in the Chugoku zone, 1–7 August (high case)   238
12.18   Grid balance in the Kansai-Chubu zone, 1–7 August (high case)   239

Tables

0.1	Territorial organization in France	10
0.2	Territorial organization in Japan	11
1.1	EU energy transition targets and French energy targets	19
4.1	Izumisano investment from stakeholders	78
5.1	Comparison of the RPS and FIT systems	88
5.2	Weighted values of REC for various NRE sources	89
5.3	Chronology	92
7.1	List of new municipal power suppliers	125
10.1	Key financial data: the case of Jühnde	187
12.1	The positioning of nuclear and renewable energies in the main energy plans	217
12.2	Regulatory framework of renewable energies and grid rules for power grid systems	218
12.3	Grid voltage levels and renewable energy integrations in TSO zones	222
12.4	Grid capacity of cross-regional interconnections in 2020 and COOTO's grid development plan for 2029	224
12.5	Target capacity of the high case	229
12.6	Targets of the high case in the Kyushu zone	230
12.7	Simulation results in the Kyushu zone (May)	237
12.8	Simulation results in the Chugoku zone (May)	237
12.9	Simulation results in the Shikoku zone (May)	237
12.10	Simulation results in the Kansai-Chubu zone (May)	237
12.11	Simulation results in the Kyushu zone (Aug)	240
12.12	Simulation results in the Chugoku zone (Aug)	240
12.13	Simulation results in the Shikoku zone (Aug)	240
12.14	Simulation results in the Kansai-Chubu zone (Aug) with additional demand decrease by 15%	240

Contributors

Luis Roman Arciniega Gil, lecturer and researcher in sustainability, Lille Catholic University. PhD candidate in Law, Lille University, CNRS - Centre for Administrative, Political and Social Studies and Research (UMR 8026), Lille, France.

Guillaume Dezobry, Senior Counsel lawyer at FIDAL and Associate Professor at Amiens University, Amiens, France.

Laure Dobigny, Postdoctoral Researcher in Socio-anthropology and Lecturer at ETHICS - EA 7446, Catholic University of Lille, and Associate Researcher at Centre for Administrative, Political and Social Studies and Research (UMR 8026), Lille, France.

Magali Dreyfus, CNRS Researcher in Law, Centre for Administrative, Political and Social Studies and Research (UMR 8026), Lille, France.

Kenji Inagaki, Executive Director, "Local Good" General Incorporated Association, Tokyo, Japan.

Masayoshi Iyoda, Research Fellow at Kiko Network, Lecturer at Graduate School of Kyoto Women's University, Japan.

Bomi Kim, Manager, ESG Planning Team, ESG New Deal Planning Department, KDB Bank, Seoul, South Korea.

Raphaël Languillon, Senior researcher – French Research Institute on Japan, Tokyo, Japan.

Youhyun Lee, Assistant Professor, Department of Public Administration, Ajou University, Suwon, South Korea.

Blanche Lormeteau, CNRS Researcher in Law, Western Institute of Law and Europe (IODE), Rennes, France.

Takuo Nakayama, Senior Lecturer, Chiba University of Commerce, Chiba, Japan.

François-Mathieu Poupeau, Senior Researcher in Political Science at CNRS, "Research Centre on Technologies, Territories and Societies" (LATTS), Paris, France.

Aki Suwa, Professor, Kyoto Women's University, Kyoto, Japan.

Asami Takehama, Professor, Industrial Sociology, Ritsumeikan University, Kyoto, Japan.

Manabu Utagawa, Senior Researcher in Energy Sciences, Sustainability and System Analysis Research Group, National Institute of Advanced Industrial Science and Technology (AIST), Tsukuba, Japan.

Marie-Anne Vanneaux, Associate Professor in Public Law, Artois University, Douai, France.

Acknowledgments

This book is based on discussions held during the three-day conference "Local energy governance in France and Japan", held at Lille University, on 11–13 September 2019. The event was mostly funded by a French National Centre for Scientific Research (CNRS) and the Japan Society for the Promotion of Science (JSPS) fellowship for the Bilateral Joint Seminars (2019). CERAPS (Centre for Administrative, Political and Social Studies and Research – UMR 8026) and Lille University also supported the event.

Introduction

Magali Dreyfus and Aki Suwa

Context

Energy, in many ways, is probably the biggest challenge of the 21st century. It bears economic, social and environmental implications from the global level to local territories. Energy supply has always been a major challenge for human societies in order to satisfy their needs from the most basic to the most sophisticated. In that regard, the rise and exploitation of fossil fuel energy sources during the industrial revolution, at the end of the 19th century, has been a major historical step that also marked the beginning of humanity's struggle to cope with its sustainability side-effects. Indeed, as energy demand has reached unprecedented levels, our economic model also seems to attain its environmental limits.

Already in 1972, with the first Club of Rome report "Limits to Growth", the problems of natural resource depletion and environmental degradation, in an ever more populated world, were identified as a potential cause of the global economic crisis. Twenty years later, the United Nations Framework Convention on Climate Change (UNFCCC) acknowledged the risks that global climatic change posed on our societies. Since then, Parties to the Convention committed to reducing their greenhouse gas (GHG) emissions associated with the use of fossil energies. As the 2015 Paris Agreement concluded, human societies agreed to limit global warming to 1.5°C to avoid the direct consequences of climate change (IPCC, 2018). Energy production generates 80% of GHG, thus the energy system transformation is imperative to global sustainability.

On a similar track, the Sustainable Development Goals (SDGs), adopted by the General Assembly of the UN, call states to ensure access to affordable, reliable, sustainable and modern energy for all. Against these international backgrounds, international and domestic energy systems need to be adjusted to the framework capable of increasing the share of renewable energy in the global energy mix, as well as achieving energy efficiency. To do so, the High Level Political Forum following the implementation of the SDGs adopted a Global Action Plan in 2019 for decentralized renewable energy. Four priorities were put forward: adopting supportive policy and regulation; unlocking finance; working in a multi-stakeholder approach; and strengthening the role of the people.[1] This aligns with the 2017 statement from the International Energy Agency, which foresees that by

DOI: 10.4324/9781003025962-1

2030, more than 71% of new electricity connections will be via off-grid or mini-grid solutions (Berka and Dreyfus, 2021).

Against this background, many states are reviewing their energy policies in order to foster the development of renewable energy. Yet this often questions the way the whole sector is organized. For instance, for matters of economy of scale, the deployment of nuclear power requires a very centralized and upper-scale organization. Fossil fuels also led to the constitution of big corporations. Yet renewable energies are based on local physical contexts (amount of sunshine, wind intensity, the strength of a river stream, etc.) and as such have a strong territorial identity. Their implementation, therefore, requires information, consultation and often authorization from local actors, be they local governments or residents.

The need for reaching sustainability in the energy market is a growing concern in countries that have long histories of centralized energy markets with strong energy industrial players. With the imperative to attain ambitious targets regarding the reduction of GHG emissions and more renewable energy, these countries face a potent challenge to reconcile conflicting positions. In order to incubate the required energy transition, comprehensive governance schemes, which integrate legal, sociological and technological domains, are required. Yet to date, energy production systems in many countries have their origins in the era of the industrial revolution, where the norm was categorically defined by the heavy reliance on fossil fuels (and later nuclear use).

Defining local energy governance: Why local energy governance is important

The challenges are, however, not only at the national level but also at the local level, which in our view, refers to an organization based on a proximity principle. In fact, environmental concerns drive the motivation to shift away from centralized energy production with fossil and nuclear toward more decentralized production, distribution and consumption of locally available renewable sources.

The spectrum of participants to distributed energy production and consumption, in contrast to the centralized model, is inevitably wide. Stakeholders for a decentralized energy project, for example, could include highly trained experts, but also non-technical citizens who may happen to live in the vicinity of the development site. With the diverse and potentially conflicting social settings, local governments and actors, such as communities, citizens and local companies, are important players who can offer innovative and flexible solutions.

There is therefore a need to analyze if and how the energy transition is taking place at the local level, with a particular focus on the emergence and structure of sociotechnical governance systems. Several papers have already studied institutional frameworks underlying the weight of relations between different local actors and identified barriers, as well as drivers of action (Kousky and Schneider, 2003). Most of them describe the complex multi-level framework within which local authorities take action, insisting on the importance of coordination between actors and policy instruments for success (Collier, 1997; Schreurs, 2008; Bulkeley and Betsill, 2003; Bulkeley and Betsill, 2013; Ostrom, 2014; Yalcin and Lefèvre, 2012; Zeemering, 2012).

More recently, academics started to integrate and dwell on the environmental or climate change framing of local public policies. Some authors have shown how the traditional centralized energy system is not suitable for the new global challenges, calling for a decentralization of the activities, thus creating a new model of governance (Eyre, 2013). In this discourse, "community" energy development is often addressed in relation to public purposes and common welfare within scientific and technological development (Jasanoff and Kim, 2015).

In fact, the transition to community energy development cannot be based only on a technical shift: redistribution of powers and competencies across a wide spectrum of stakeholders is key to the efficacy and effectiveness of low carbon community energy elaboration (Bridge et al., 2013). However, despite the emergence of this significant discourse, there is room for more research to be done, especially on the governance system that may enable the redistribution of powers and competencies. Without analyzing this, it becomes highly difficult to materialize community energy. It is worth noting that although in these studies legal and institutional framework is often mentioned as an important determinant in the system (Schroeder and Bulkeley, 2008), especially with regard to the autonomy of local governments, specific governance studies, focusing on local authorities' actions remain scarce (Richardson, 2012; Sperling, Hvelplund and Vad Mathiesen, 2011; Engel, 2006).

In this literature, there is no fixed definition of what local energy governance is. Wade et al. (2013), define energy governance as "the rules, processes, practices and behaviour that affect the way in which energy is generated and used" (Wade et al., 2013). This underlines the strong institutional dimension of governance, and the need to look over different domains: law, politics, economics, environmental sciences and so on. As for local, which is our key feature of interest, it may represent a level of government, but also a specific territory marked by specific physical, cultural, social and economic characteristics, which constitute its proper identity. We, therefore, in the context of this book, study decentralized production, distribution and consumption of locally available renewable sources and the different actors involved in these processes.

It is also important to recall that at the local level there are many other key actors in the energy sector: the central government administration, private commercial entities, big energy market players and incumbents and so on, whose interactions with local actors need to be studied. Special issues about these topics are increasing with numerous case studies. This reveals that the understanding of these concepts, their scope and the agents and forms of decentralization are country- and context-specific (Berka and Dreyfus, 2021).[2] Against this backdrop, this book aims to contribute to these studies by providing a comparison of the development of local energy governance in France and Japan.

Why France and Japan?

The change in energy sources comes along with a change in governance. Energy networks are sociotechnical systems, which are based on long-term institutions that create path-dependency. "Local energy governance", therefore, is easy said but difficult to materialize.

Economic and political justification often explains the formation of a centralized energy system. In France, a state-wide electric utility, *Electricité de France* (EDF), was created after World War II in order to boost energy independence and security of supply. It was in a monopolistic situation in all segments of the energy chain. Prior to 1946, there was a significant number of local companies in France (see Chapters 3 and 6). The industrial base of EDF was initially developed with fossil-fired generation facilities using coal, oil and hydropower, and then from the 1960s with nuclear plants. The French model was organized around agreements that give access to electricity at its production cost, until later when market liberalization disconnected economic association through the introduction of wholesale market prices (Reverdy, 2015). Lately, the major development in France has been the opening of the market under the influence of European regulations (see Chapter 1). This reform has been met with a strong reaction from the industrial coalition, a coalition that has been exercising strong political influence over the electricity market. Thus, the entry of such new local actors, especially renewable based, faces tension with incumbent national players.

In Japan, the historical context demonstrates a certain similarity with the French case. In the early days of the Japanese modernization period of the late 19th century, more than 30 electric companies were established throughout the nation. The early 20th century marked the establishment of 700 electric companies, which merged to create five major electric companies after the First World War. During the Second World War, the electric utility industry was completely state-controlled, and these companies were integrated into Nihon Hatsusoden Co. (a nationwide power generating and transmitting state-owned company) and nine distribution companies to establish the war-time resource consolidation.

After the end of the Second World War in 1945, under the influence of US policy, regional privately owned and managed electric utilities – Hokkaido, Tohoku, Tokyo, Chubu, Hokuriku, Kansai, Chugoku, Shikoku and Kyushu Electric power companies – were established in 1951 (and later Okinawa Electric Power Company joined in this list after the return of Okinawa from US occupation in 1972). Though these have the form of private companies, they are a vertical monopoly in the respective region (FEPC, 2021).

At the end of the 20th century, mainly an international demand to push down electricity prices in Japan formed a sufficient political drive to deregulate the Japanese electric market. In December 1995, independent power producers (IPP) were allowed to carry out wholesale sales, and in March 2000, the electricity retail supply for extra-high voltage users (demand exceeding 2 MW) was liberalized. The scope of retail liberalization was then expanded in April 2004 to users of more than 500 kW, and subsequently in April 2005 to users of more than 50 kW.

The Fukushima Daiichi Nuclear Power Station accident, following the Great East Japan Earthquake in March 2011, triggered much debate as to the problems associated with the nation's nuclear development policy (Chapter 4). One of the issues was the "non-transparent" financial system that allows the utility to invest (often too much) into nuclear technology and development. Since 1933, the "total cost method" was introduced, which became the basis for electricity tariff calculation. The total cost method is to determine the tariff by adding a rate of return on top of the total cost the utility invests. As the balance between cost and income is ultimately secured, it functions as a mechanism for Japanese governments to support electric utilities with long-term capital investment, including nuclear plants.

The total cost method was heavily criticized, especially after 2011, because it had become the fundamental engine to encourage utilities to over-invest in capital-intensive purposes, without sufficient transparency in cost accountability. Reflecting on this criticism and the punitive gesture to stop utilities' nuclear over-investment, full retail liberalization finally started in April 2016, marking the end of the total cost method that conflicts with market competition policy.

The liberalization of energy markets in many industrialized countries has opened space for new entrants that are able to challenge, to a certain extent, the main incumbents and offer alternative energy services. In France and Japan, two very centralized countries, the government is the main actor, along with a national monopoly operator in France and nine regional monopolies operators in Japan. Over the past decades, as seen earlier, France and Japan both relied substantially on nuclear power to provide electricity efficiently to their entire territory at affordable prices. France is the second-highest nuclear plant user in the world, after the United States, while Japan used to be the third. Yet the Fukushima disaster of 2011 called into question that source of energy in Japan and abroad (e.g., Germany).

Although the first goal of these liberalization processes was to boost competition among utilities in order to reduce energy prices for costumers, it has now been combined with ambitious targets in terms of GHG emission reduction, the share of renewable energy in final energy consumption and energy efficiency.

In France, in 2015, the Energy Transition for Green Growth law set the goal of reducing the share of nuclear power in energy consumption from 75% of the electricity mix to 50% in 2025, yet this was postponed to 2030. In Japan, the first reaction to the Fukushima disaster was to abandon that source of energy. Yet, slowly, reactors were restarted (nine in 2021). Moreover, the 2015 Paris Agreement on Climate Change provides that parties should reduce their GHG emissions to limit the global temperature rise this century to below 2°C above pre-industrial levels. Japan committed to reducing GHGs to net-zero by 2050 and France aims for a 75% reduction by 2050 with respect to 1990. Yet the common feature of the two energy systems, following a global trend, is that they are now both considering increasing the share of renewable energy in their energy mix. France's goal is to reach 33% of renewable energy sources (RES) in gross final consumption by 2030, and, by 2030, reach a mix of 40% renewables in electricity production. The Japanese goal is to achieve 36% to 38% of renewables in electricity production by

2030, although it has been reported that Japan has already achieved 18% (including large hydropower) in 2019.

With the support of renewable energy, new forms of governance are now being promoted. Emerging in this renewed policy context, new entrants will therefore be key players in pursuing these goals. In this book, we focus on these new actors at the local level, and on local authorities, communities and citizens, individually and collectively, who in addition to economic and environmental concerns, often put a strong emphasis on other social and political dimensions, such as justice or democratic issues. Since local energy projects, initiated by local governments, citizens and communities, or private stakeholders, are now increasing in France and Japan, these two countries provide a rich field of study, with conclusions that will be informative for other countries aiming at promoting renewable energy. Also, to enrich these studies, a look at the status of local energy governance and renewable energy development in third countries is also provided (Austria and Germany, South Korea, Mexico).

Questions and chapters in the book

The underpinning questions of the different chapters are: How do central governments promote local energy transition and governance by supporting local actors in this process? What circumstances trigger local authorities to take action and which benefits can they draw from it? And more broadly, is this renewed energy governance changing life (e.g., the perception of wellbeing) in local areas for different stakeholders, and what are the benefits? Are the new schemes of energy management more advanced and innovative than at the national and/or global level?

To answer these questions, the different chapters cover several academic disciplines, proving a multi-disciplinary perspective on the issue of local energy governance development.

"Local Energy Governance: Opportunities and Challenges for Renewable and Decentralised Energy in France and Japan" pays particular attention to community renewable policy development and implementation in France and Japan, which have rapidly changing regulatory and institutional contexts.

Part I outlines the national institutional frameworks (rules and main actors) within which local energy governance is developing in France and Japan, laying down the state of the debate, as well as highlighting how much they are influenced by the international (and European) context.

Chapter 1 provides an overview of the French electricity sector and the predominance of nuclear power in the energy mix and in the regulatory framework, which is largely organized around that source of energy. It also highlights the key role of the European Union in placing the climate change issue in the national political agenda, especially by fixing reduction targets for GHG emissions as well as the share of renewable energy in the final consumption – later transposed by French legislation. Interestingly enough, Guillaume Dezobry and Magali Dreyfus show that it is through the same channels that local governance in the energy sector emerged. As a result, today, the French legal framework promotes decentralization as well as self-production and self-consumption of renewable energy, thus breaking with the traditional centralized model and showing an unprecedented openness toward new actors. However, its implementation still needs to be verified.

Chapter 2 deals with the political landscape and civil society's advocacy toward 100% renewables in Japan. The Japanese government, tied with Japan's industry groups, traditionally favored conventional energy such as coal and nuclear rather than renewable energy. Nevertheless, civil society became very active in the search for a "just transition to 100% renewables" to address the climate crisis and nuclear disasters. Masayoshi Iyoda describes how civil society strives to promote such a transition through advocacy and lobbying, community-based renewable energy projects and public campaigns. Recent policy developments indicate that this approach has had some success.

Chapters in Part II make an in-depth analysis of the policy integration and fragmentation on local government powers in the energy sector in the liberalized electricity market. We observe which powers and resources local governments have to take action in the energy sector with respect to national authorities. We observe what specific duties and powers were given to them to develop a specific local governance scheme, but also tensions between public authorities.

Chapter 3 shows that local authorities play an increasing role in the management of the electricity sector in France. Tapping opportunities open with liberalization and decentralization processes, they are involved directly in production, distribution and supply activities. They are also indirectly acting on the energy sector on the basis of their powers in the intervention in planning and urban development, housing and energy poverty, mobility and so on. François-Mathieu Poupeau thus provides an update on this growing role but also points to its limits, showing how a coalition of actors supporting the centralized model continues to exercise control over the strategic functions of the sector, thus limiting the capacity of action of local authorities.

Chapter 4 introduces the Japanese background and the central government policy context for local renewable energy development in Japan. It provides a brief historical outlook on the liberalization process and explains how energy security has always been a major concern for Japan. Through two brief case studies, Aki Suwa shows how current socio-economic conditions in rural regions, such as stringent financial restrictions due to falling birth rates, an aging population and the decline in overall population, foster the development of local energy governance through the various benefits associated with it.

Finally, Bomi Kim and Youhyun Lee provide a case study set in South Korea. They raise the fundamental question of the social acceptance of renewable energy projects. Following the chronology of the Garorim Bay tidal plant project, they show how a top-down system of decision risks facing rejection by local actors. They conclude by emphasizing the importance of collaborative governance at all stages of the project – before its approval, its design and its operation.

Part III investigates the opportunities that different local actors (local governments, communities, citizens) may find in constituting a local company aiming at providing energy services. The focus is on energy businesses and their legal forms, for example, public companies, public-private partnerships and cooperatives, as it is important to highlight how the different legal forms involve different actors, with different powers relations, which ultimately influence the shape of local governance.

Chapter 6 focuses on French local public enterprises as a tool for local governments to implement their energy policy. There are several kinds of enterprises, and the legislator has created new legal forms to foster energy transition at the local level. It highlights their missions and specificities. Through the analysis of the shareholding composition of these enterprises, Marie-Anne Vanneaux explains the circumstances in which a local government will favor one legal form over another. Financial needs, technical expertise requirements, acceptance through participation and a desire to remain the main decision-maker in the local policy, in addition to legal requirements, are among the variables that guide the final choice. Thus, energy transition and development of renewable energy is not always the most decisive factor.

In Chapter 7, Kenji Inagaki and Takuo Nakayama present an economic assessment of local renewable projects. They show how municipal power suppliers have recently found success in Japan, despite a very traditional centralized organization of the energy sector. The economic benefits appear to be a real incentive for local authorities to engage in the sector. The chapter highlights the value added to local economies, and how the values can be changed depending on the business structure of local power suppliers.

Part IV argues whether and how local authorities can achieve 100% renewable territories. In that regard, local governments have adopted this ambitious target before their national governments. The ambition to become more resilient and autonomous is often the backbone of that goal. This objective is often reached through strategies that combine renewable energy generation, energy efficiency and the reduction of energy consumption (energy conservation or "sobriété"). In this part, the authors explore the territories' motives, incentives and barriers to their actions.

Chapter 8 focuses on bottom-up initiatives, where "territories" (meaning local governments with other local public and private actors) have the ambition to become resilient to different kinds of stresses (environmental, health, industrial, etc.) through the achievement of energy autonomy. The French "TEPOS", which aimed at producing enough renewable energy to meet local demand, pursued this ambition and were successful enough to attract central government's attention. Yet, Blanche Lormeteau shows how when the TEPOS initiative was taken over by national authorities, it failed to maintain some of its best assets, namely its flexibility and its tailored features to the specific territory where it was deployed.

Chapter 10 explores other international experiences through the examination of four case studies in Austria and Germany, two federal countries (contrary to France and Japan). There, citizens' participation and empowerment have a long history. This socio-anthropological study in rural areas, led by Laure Dobigny, helps to understand the motives and social implications of 100% renewable energy territories. Despite the different contexts, the chapter shows commonalities with France and Japan in local concerns (energy security, economic development, justice, etc.) and emphasizes the various values at stake in local energy governance.

Part V explores technological and digital issues in energy transition. Online access to the electricity market, ancillary control and the relevant information and communication technologies seem to be the new imperatives of energy transition. This final part questions how infrastructure, such as grids, should be managed to promote renewable energy development, as well as the limits of technological approaches.

In Chapter 11, Raphaël Languillon-Aussel sets out the linkages between the popular concept of a "smart city" and energy governance in France and Japan. The smart city model has met great recognition worldwide and may in fact help, through information and communication technologies, in curbing energy consumption and adjusting the production of renewable energy to energy demand. Yet as the chapter shows, despite a name that evokes the urban level of government, little room is left to local actors. Solutions are discussed at the national level and appear imported, often by big technological corporations. Local actors, in particular citizens, are left aside, which impedes any contextualized solution and any real local governance.

Chapter 12 presents the current situation of the energy mix in Japan, some regulatory deadlocks in the introduction of renewable energy, and the lack of cooperation from the main incumbents. The authors, therefore, question the feasibility of a renewable electricity target of 40% by 2030 in the western Japan grids by utilizing a simplified unit commitment model (UCM). This would partly help Japan meet its national renewable targets. Using the UCM, Asami Takehama and Manabu Utagawa show how the deployment of cross-regional interconnections, heat pumps and electric vehicle units could boost more renewable energy in the studied area.

The last chapter of the book examines how local energy governance develops in Mexico. It does so through a critical approach to the concept of a smart city, which has gained recognition in Mexico. By reconsidering what is "smart", Luis Román Arciniega Gil demonstrates how any technological approach must be combined with social considerations such as fighting energy poverty, and self-production of energy within the industrial and business sector especially at the local level. Thus, local renewable energy projects appear like a good tool to fight poverty at large, in line with the SDGs.

Finally, to help the readers, two tables summarize the political organization of the two countries, both being unitary countries but with some degree of decentralization (see Table 0.1 for France and Table 0.2 for Japan). The energy mixes are recapped in Chapters 1 and 12 for France and Japan, respectively.

Table 0.1 Territorial organization in France

France is characterized by a great complexity in the organization of local and regional authorities. In this book, five main levels are mentioned.

The oldest level of government (since the Middle Ages) is the *communes* **or municipalities**. There are about 35,500. Their remit and powers have varied according to the period but the laws of decentralization (from the beginning of the 1980s) have considerably strengthened them in terms of town planning, roads, transport, social action and culture, to name but a few areas. Municipalities have sometimes managed some of their public services in an intermunicipal form by setting up **unions of municipalities** (*syndicats de communes*) in charge of managing energy, water, or other activities, which they considered more efficient to entrust to structures with a broader scope. There are currently more than 10,000 unions with approximately 80 different jurisdictions[a]. Not all of them are simply governed by municipalities, as other levels of local government may be involved.

Above municipalities are the *départements*. They date back to the French revolution and are currently 94 local governments at the level of *départements*. They have strong competences, particularly in road management and social action.

New forms of **intercommunalities or municipal groupings** (*intercommunalités*) appeared in France during the 20th century. These are structures that, unlike the unions of municipalities, are organized to manage several areas at the scale of a territory that is supposed to make sense from an economic (job pool) and social (life pool) point of view. Intercommunalities began to develop in France from 1959 onward in various forms. Currently, four main types can be distinguished (in order of importance in terms of population and powers exercised): *communautés de communes*, *communautés d'agglomération*, *communautés urbaines* (for big cities theoretically) and *métropoles* (metropolises). In particular, the metropolises, created from 2010 onward, are entrusted with increasingly important competences, including, recently, energy.

Régions were established in France in 1972. Their representatives have been elected by universal suffrage since 1986. They have an increasing number of responsibilities, as public authorities consider that, because of their scope and size, they are best placed to deal with issues such as economic development, transport and land planning. There are currently 18 (including 13 in metropolitan France), following mergers carried out in 2015.

---

[a] Figures of the Ministry of the Interior.
(Source: François-Mathieu Poupeau, see also Chapter 3.)

Table 0.2 Territorial organization in Japan

Japan's local autonomy system adopts a two-tier system of prefectures as regional government units and municipalities as local government units. Japan's current local autonomy is based on the Constitution of Japan, which was adopted in 1946 and took effect the following year, but its origin dates back to the early 19th century when the Meiji Reformist government adopted the system of Prussia, which ensures strong centralized power to be exercised at the regional and local government levels, to catch up with Western countries.

Although prefecture and municipal governments in Japan have various populations and sizes, they are all basically given similar powers to conduct their duties.

Thus, local governments in Japan demonstrate uniform characteristics in both their organization and their administrative operations. This is also generally supported by the public expectation that services provided by local governments should be uniform in their contents and standards irrespective of the region of the country. In order to equalize local governments in different contexts, the local allocation tax system has been established that distributes a prescribed portion of the national tax to local governments according to their financial power. It is at the same time controversial as to what extent the local allocation tax system became the tool for the national government to take control of local tiers.

Under the current Local Autonomy Law, local governments are classified into two types: ordinary local public entities and special local public entities. Prefectures and municipalities are ordinary local public entities.

Prefectures

The Japanese terms Do, Fu and Ken are all rendered as "prefecture" in English. In Japan, there are currently 46 of these wide-area local governments encompassing municipalities (47 including Tokyo). The term To is only applied to the Tokyo Metropolitan Government, although it technically designates a prefecture. The main duties of prefectural government include prefectural infrastructural construction and maintenance, forest and river conservancy, operating public health centers and vocational training and police functions.

Municipalities

Municipalities constitute the basic level of government handling issues closest to the lives of residents. Duties include resident registration and issues related to the safety and health of residents (e.g., fire service, garbage disposal, water supply, sewage), the welfare of residents (e.g., public assistance, nursing insurance, national health insurance) and urban development plans (e.g., urban design, municipal roads, parks), among others.

Prefectures and municipalities are mutually independent and there is no official hierarchical relationship between them. However, prefectures often give guidance and advice to municipalities, also performing various licensing and permit functions. Also, it is worth mentioning the importance of decentralization, which has been increasingly recognized in Japan, and there have been various legal changes in recent years.

(Source: Council of Local Authorities for International Relations, 2010.)

Notes

1 See https://sustainabledevelopment.un.org/content/documents/24075ab5_cover.pdf.
2 See, among others, Muluguetta J., Jackson T., van der horst, D., 2010, "Carbon reduction at community scale", *Energy policy*; Correlje A., Hoppe T., Künneke R., 2019, "Sustainable urban energy systems: governance and citizen involvement", *Energy policy*; Dreyfus M. et Berka A., 2020. Special issue on "Energy Decentralization: institutional perspectives", *Renewable and Sustainable Energy Reviews (RSER);* Koirala B., Parra D., Bauwens T., 2021, "Energy Communities in the Changing Energy Landscape", *Sustainability.*

References

Berka, A. and Dreyfus, M. (2021) 'Decentralisation and inclusivity in the energy sector: Preconditions, impacts and avenues for further research', RSER, vol. 138, in Dreyfus, M. and Berka, A. (eds) *Renewable and sustainable energy reviews (RSER).* Special issue on 'Energy decentralization: Institutional perspectives', *RSER,* 2020–2021.

Bridge, G., Bouzarovski, S., Bradshaw, M. and Eyre, N. (2013) 'Geographies of energy transition: Space, place and the low-carbon economy', *Energy Policy,* 53(C), pp. 331–340.

Bulkeley, H. and Betsill, M. (2003) 'Cities and climate change: Urban sustainability and global environmental governance', *Annual Review of Environmental Resources,* Vol. 35:229-253, pp. 229–253.

Bulkeley, H. and Betsill, M. (2013) 'Revisiting the urban politics of climate change', *Environmental Politics,* 22, pp. 136–154.

Collier, U. (1997) 'Local authorities and climate protection in the European union: Putting subsidiarity into practice?', *Local Environment,* 2–1, pp. 38–57.

Council of Local Authorities for International Relation (2010) 'Local government in Japan', http://www.clair.or.jp/j/forum/series/pdf/j05-e.pdf

Engel, K. (2006) 'State and local climate change initiatives: What is motivating state and local governments to address a global problem and what does this say about federalism and environmental law?', *Urban Law,* 38, p. 1015.

Eyre, N. (2013) 'Decentralisation of governance in the low carbon transition', in Fouquet, R. (ed.) *Handbook of energy and climate change.* London: pp. 581–597.

Federation of Electric Companies of Japan (FEPC), 'History of Japan's electric power industry 2', https://www.fepc.or.jp/english/energy_electricity/history/ (accessed on 18 August 2020).

IPCC (2018) 'IPCC special report on global warming of 1.5', https://www.ipcc.ch/sr15/

Jasanoff, S. and Kim, S.-H. (eds) (2015) *Dreamscapes of modernity: Sociotechnical imaginaries and the fabrication of power.* University of Chicago Press, Chicago.

Kousky, C. and Schneider, S.H. (2003) 'Global climate policy: Will cities lead the way?', *Climate Policy,* 3, pp. 359–372.

Ostrom, E. (2014) 'A polycentric approach for coping with climate change', *Annals of Economics and Finance,* 15–1, pp. 71–108.

Reverdy, T. (2015) 'Managing uncertain reform through "flexible institution": Electricity sector liberalization in France', in *SASE Conference: Inequality in the 21st Century,* July 2015, London.

Richardson, B. (2012) *Local climate change law environmental regulation in cities and other localities.* New York: IUCN Academy of Environmental Law Series.

Schreurs, M. (2008) 'From the bottom up: Local and subnational climate change politics', *The Journal of Environment Development*, 17(4), pp. 343–355.

Schroeder, H. and Bulkeley, H. (2008) 'Global cities and the governance of climate change: What is the role of law in cities?', *Fordham Urban Law Journal*, 36, p. 313.

Sperling, K., Hvelplund, F. and Vad Mathiesen, B. (2011) 'Centralisation and decentralisation in strategic municipal energy planning in Denmark', *Energy Policy*, 39, pp. 1338–1351.

Wade, J., Eyre, N., Parag, Y. and Hamilton, J. (2013) 'Local energy governance: Communities and energy efficiency policy', in *Proceedings of the ECEEE 2013 Summer Study on Energy Efficiency*, Stockholm, Sweden, 3–8 June 2013, pp. 637–648.

Yalcin, M. and Lefèvre, B. (2012) 'Local climate action plans in France: Emergence, limitations and conditions for success', *Environmental Policy and Governance*, 22–2, pp. 104–115.

Zeemering, E. (2012) 'Recognising interdependence and defining multi-level governance in city sustainability plans', *Local Environment*, 17(4), pp. 409–424.

Part I

National Framework of Energy Governance

1 Searching for alternatives to fossil- and fission-based energy sources in France

Guillaume Dezobry and Magali Dreyfus

In the years 2000, the rising of the climate issue as a priority on political agendas has impacted European and national energy policies. One of the most direct consequences is the search for clean energy sources, meaning low carbon, which translates into the promotion of renewable energies. Yet, in France, where electricity mostly comes from nuclear power, considered a low greenhouse gas (GHG) emitting source, the move has been slow. Urged by the European Commission and a binding European Union (EU) regulatory framework, a change is underway. This chapter addresses the question of potential deadlocks regarding energy transition (understood as a change of energy source) in the institutional setting. It recalls how EU law boosts change, but it is faced with a socio-technical giant, namely the nuclear industry, which finds in economics theory a strong legitimacy to pursue its activity. However, current changes in national law open the door to a new governance scheme, focusing on local actors. Its effectiveness is still to be tested.

The European energy and climate institutional framework

To understand the French energy and climate institutional framework, it is key to recall the European regulatory framework, which is binding for French authorities. In fact, since its foundation, the EU has dealt with energy issues.[1] Yet over the years, the different rules became increasingly ambitious in their scope, and technical and intrusive in the EU member states' legal orders. The recent "Clean Energy for all Europeans Package", adopted in the aftermath of the 2015 Climate Paris Agreement, is the most recent comprehensive set of reforms enacted by EU institutions. It covers a wide range of topics from the development of the European energy single market to the promotion of renewable energy.[2] In this section, two points particularly relevant to the topic of that book are detailed: the liberalization of energy markets and the promotion of renewable energy to tackle the issue of climate change.

The liberalization of energy markets

The liberalization of energy markets (basically in the electricity and gas sectors) started in the 1990s with the aim of contributing to the achievement of the

DOI: 10.4324/9781003025962-3

European single market and offering cheaper energy to European consumers. The whole process consisted in introducing competition in order to allow new operators to penetrate the market. This clashed with the French model where the main incumbent 'Electricité de France' (EDF), a public company largely owned by the state, was in the situation of being a monopoly. Thus, EDF, as a vertically integrated company, managed the whole chain of activities from generation to supply to final users. In the 1990s, the EU adopted various directives[3] that started unbundling the energy sector and opened it to competition generation and supply to final users, while transmission and distribution remained regulated in a *de facto* situation of monopoly.

As shown in Figure 1.1, the French energy market remains dominated by one incumbent, the former EDF, which split into several branches, and where the state is the main stakeholder. More than 20 years after the first liberalization wave, the French energy sector is therefore still quite closed and centralized, with an incumbent where the main activities lie in the nuclear industry.

The promotion of renewable energy and energy transition

Coupled with the security of supply, climate change is a key topic in European politics. Mitigating it and adapting to it have become major issues in European public policies as the EU developed important and very visible diplomacy around this question in the international arena. To a certain extent, this bound it to become exemplary. As a result, its action is oriented toward three main goals: reducing GHG emissions, increasing energy efficiency and raising renewable energy production and consumption. Here we focus on the latter dimension of the EU's activity.

The promotion of renewable energy started in the 2000s and has led to fostering the development of generation units that supply local consumption areas. Thus, local governments gained renewed attention as key actors of the sector. The major benefits are a better security supply, shorter transmission distances and a reduction of the losses associated with energy transmission. Table 1.1 shows the different ambitious targets that the EU is aiming at, as well as the French ones.

Member states are free to determine how to reach these goals and as a principle, the EU does not intervene in the internal administrative organization of its member states. It only has an economic approach of the different players in public policies. It means that it deals with public actors, such as local authorities, only as market actors. However, by creating legal categories for the energy market, the EU extends the scope of intervention of local governments and civil society.

Figure 1.1 Liberalization of the electricity market in (metropolitan) France.

Table 1.1 EU energy transition targets and French energy targets

	EU goals according to the Clean Energy Package for all Europeans	*French self-assigned goals to contribute to reach the European goals (Art. L. 100-4 Energy Code)*
GHG reduction	Reduction of 40% by 2030 from 1990 levels (binding goal)	Reduce GHG emissions by 40% between 1990 and 2030 (Energy Code as amended by the law on Energy and Climate)
Promotion of renewable energy	Reach 32% of renewable energy sources (RES) in the EU's energy (gross final consumption) mix by 2030 (binding goal)	Reach 23% of RES in gross final consumption in 2020 and up to 33% by 2030; by 2030, reach a mix of 40% renewables in electricity production, 38% in heat final consumption and 10% of gas consumption
Energy efficiency	Target of 32.5% for energy efficiency for 2030, compared with a baseline scenario established in 2007 (non-binding goal – best efforts obligation)	Reduce by 50% the final energy consumption by 2050 with respect to 2012. By 2030, reduce 2030 the primary fossil fuel energy consumption by 40%
Low carbon economy	Climate-neutral economy by 2050 (European Green Deal adopted in 2019)	By 2050, achieve carbon neutrality throughout the country, without the use of carbon offsetting, by reducing gross emissions by a factor of at least six compared with 1990

Source: the authors.

Acknowledgment and promotion of new actors in the European local energy governance

Lately, European energy law, following a trend in different member states, has acknowledged and started promoting new local actors to foster an energy transition to more renewable sources. In the Clean Energy Package, Directive 2019/944[4] provides that final customers should be allowed to sell on the market their own production of electricity. In addition, Directive 2018/2001[5] from the same package recognizes the right to self-consumption of electricity, individually or collectively. As regards, collective self-consumption, European law created two legal categories, namely "Citizen Energy Communities" (CECs) (Dir. 2019/944) and "Renewable Energy Communities" (RECs) (Dir. 2018/2011), which still have to show their operationality (Lowitzsch, Hoicka and van Tulder, 2021; Heldeweg and Saintier, 2020). According to art. 2 (11) Dir. 2019/944:

"citizen energy community" means a legal entity that: (a) is based on voluntary and open participation and is effectively controlled by members or shareholders that are natural persons, local authorities, including municipalities, or small enterprises; (b) has for its primary purpose to provide environmental, economic or social community benefits to its members or shareholders or to the local areas where it operates rather than to generate financial profits; and (c) may engage in generation, including from renewable sources, distribution, supply, consumption, aggregation, energy storage, energy efficiency services or charging services for electric vehicles or provide other energy services to its members or shareholders.

Art. 2 (16) of Dir. 2019/944 provides that "renewable energy community" means

a legal entity: (a) which, in accordance with the applicable national law, is based on open and voluntary participation, is autonomous, and is effectively controlled by shareholders or members that are located in the proximity of the renewable energy projects that are owned and developed by that legal entity; (b) the shareholders or members of which are natural persons, SMEs or local authorities, including municipalities; (c) the primary purpose of which is to provide environmental, economic or social community benefits for its shareholders or members or for the local areas where it operates, rather than financial profits.

CECs and RECs are therefore non-commercial market actors, who are entitled to have non-discriminatory access to energy infrastructures. Their activities range from production to consumption, storage and selling. Only CECs though can develop their activity in distribution. Interestingly enough, the primary purpose of these actors is not profit-making, but the benefits to local communities from a social, economic, or environmental point-of-view. The promotion of that kind of actors by the EU, therefore, awards a political dimension to the energy transition process, backing the idea that local management is an ideal scale of operation.

European law with which French energy law has to comply with, has therefore constrained to open electricity markets to new entrants in France. At first, this aimed at allowing commercial actors to penetrate the markets, but lately, this also relates to local non-commercial actors such as local public authorities or even citizens on an individual or collective basis. This highlights the close link between renewable energy and their local settlements, allowing consideration of the benefits for the territories, which are not merely economic or environmental, but also social. This opening is not easy in a country such as France where the main incumbent has very close institutional ties with the French state's authorities and continues to be predominant in the market with its associated energy source, nuclear power.

An institutional framework designed for centralized governance and nuclear hegemony in the French energy mix

This section focuses on the economic regulatory framework of the incumbent French nuclear power plants. There are 58 nuclear reactors spread across 19 power plants located throughout (metropolitan) France. With an installed capacity of roughly 63 GW and an average annual production of nearly 400 TWh, the French nuclear reactors – the second highest in the world – can generate over 70% of the electricity fed into the grid.

The competitiveness of the French nuclear industry results from a combination of three factors:

1. Standardization of reactors: the 58 reactors are pressurized water reactors (PWRs). Developed in successive stages, the reactors are nonetheless very homogeneous in design and were built under very similar industrial circumstances. This standardization has made it possible to reap significant learning and pooling effects at every phase of a plant's life cycle.
2. Creation of the reactors within a short period: the French government's nuclear program was rolled out in just 20 years (from the mid-1970s to the mid-1990s). Initially designed to operate for 40 years, the nuclear plants are thus expected to reach the end of their average longevity in 2025.[6]
3. Unified management of the fleet: France's nuclear plants are run only by their incumbent operator. This choice, which was based in part on safety concerns, further enhances the economies of scale and competitiveness of this production facility.

In the context of a liberalized market, the incumbent operator's exclusive ownership of French nuclear plants was perceived, by the European authorities in particular, as an obstacle to the development of competition and as possibly constituting a "market failure",[7] particularly insofar as concerns the development of a competitive electricity supply market. The European Commission thus noted that

> In view of the scale and uniqueness of the competitive advantages conferred in the past and still conferred on this undertaking by the operation of its nuclear power capacity [...], it would have been pointless to hope that competition alone by new entrants would have permitted optimum conditions of competition to be created in the provision of electricity supply services.[8]

It was, therefore, necessary for the public authorities to intervene to correct this situation and create the conditions for workable competition in the electricity supply market.[9]

However, the selected mechanism would have to take into account two fundamental requirements for the French market. Indeed, it would have to preserve:

- on the one hand, the incumbent operator's integrated upstream/downstream character; and
- on the other hand, the integrated management of the nuclear reactors.[10]

These factors – which are key – allow us to better understand the difficulties encountered when the French market was opened up to competition and to better appreciate how important access to nuclear production is for the exercise of supply activity.

Indeed, thanks to the preservation of the incumbent operator's integrated character, the downstream branch (i.e., the supply branch) is able to purchase electricity directly from the upstream branch (i.e., the production branch, and notably nuclear production) without going through the midstream wholesale market.

In addition, the choice to preserve the unity of the nuclear plants means that the benefit of producing nuclear electricity is reserved primarily, even exclusively, for EDF. As a result, the incumbent operator's supply branch enjoys very competitive purchasing conditions that are difficult to replicate. In other words, in the absence of other means of production at comparable costs, the only way to foster the development of competition in the supply market was to allow EDF's rivals to have access to nuclear electricity at a price reflecting the costs of the nuclear power plants.

Thus, under law no. 2010-1488 of 7 December 2010 on the new organization of the electricity market (referred to as the "NOME Act"), a system of regulated access to incumbent nuclear electricity named ARENH (Accès Régulé à l'Electricité Nucléaire – Regulated Access to Nuclear Electricity) was created. The mechanism consists of granting alternative suppliers a right of access, at a regulated price, to a quantity of the electricity generated by EDF's incumbent nuclear reactors.

The mechanism is subject to two limitations:

- a temporal limitation: the mechanism expires in 2025;
- a quantitative limitation: the mechanism applies to a total maximum electricity volume of 100 TWh.

The upcoming expiration of the current system, as well as the 2019 excess of alternative supplier ARENH requests beyond the legal ceiling, bring the mechanism's limitations to the forefront and raise the question of what future regulation should be. Yet in the meantime, it is essential to understand the uniqueness of nuclear power in the French market, and how it fulfills the criteria of an essential facility.

It should first be recalled that, under established case law, a facility is "essential" to the exercise of an activity if "there [is] no real or potential substitute" for it.[11] In other words, a facility should be recognized as essential if, on the one hand, there is no way to technically or economically replace the resource in question and, on the other hand, it is impossible or especially difficult to reproduce it under reasonable economic conditions.

Assessing the essentiality of nuclear electricity involves answering the following question: is access to incumbent nuclear electricity necessary in order to operate in the French supply market? In other words, without access to that resource, is it possible for an alternative operator to compete with the incumbent operator's offer on the retail market?

The absence of any real or potential substitute in the current economic regulatory framework

This analysis involves focusing on the available means of baseload generation. Indeed, nuclear electricity is mainly a baseload electricity source (as opposed to peak-demand electricity).

Baseload electricity can be defined as "the portion of supplied electricity corresponding to the output of power plants operating continuously except for maintenance shutdown periods".[12] It is generated by production facilities whose annual use time is generally over 6,000 hours per year.[13]

From an economic standpoint, baseload generation facilities have the lowest variable costs of all so-called conventional means of production. In the French market, baseload electricity is mainly generated by nuclear power plants.

The question is therefore twofold:

- Are there any other existing means of production with the same characteristics as nuclear electricity? (Analysis of real substitutes)
- Is it possible to replicate nuclear plants? (Analysis of potential substitutes)

To analyze whether there are any existing alternative means, the focus must be placed on the two most plausible alternatives: baseload hydro-electrical plants (also called run-of-river plants) and procurement on the wholesale market.

It is clear that an alternative supplier would not be able to build a business plan rivaling EDF's offers by basing its sourcing on the wholesale market. The volatility and levels of prices observed over the past few years show that the wholesale market cannot offer a sustainable alternative to a sourcing model based on the costs of nuclear power plants.

Indeed, even though the wholesale prices for the years 2014, 2015 and 2016 were lower than the ARENH price (i.e., below €42/MWh), the (often unpredictable) price fluctuations and resulting uncertainty would make it unfeasible for such an operator to sustainably compete with the incumbent operator on the supply market.

As for run-of-river hydropower, the answer is provided by the ADLC (Autorité de la Concurrence – the French Competition Authority), which has specified that:

> the large "run-of-river" production sites, such as the dams on the Rhine or the Rhone, are not able to generate electrical power under the same conditions as nuclear plants, given the variability of the flows according to climate conditions and seasons. Even assuming a full use of these baseload generation facilities, the run-of-river capacities remain modest [...], as output is limited to roughly 8 TWh per year for the Rhine and from 12 to 15 TWh per year for the Rhone.[14]

It is equally important to point out why the other – so-called non-steerable[15] – renewable energies, whose auction prices are becoming increasingly competitive, do not seem to constitute alternatives (at least in the short and medium term). We must remember that the objective here is to identify means of production that would allow an integrated upstream/downstream operator to compete with the incumbent operator on the supply market. Under this logic, an operator would have to secure its supply purchases, particularly of baseload electricity, that is, the portion of electricity that its customers will continuously consume throughout the year. The price of electricity is therefore not the only criterion. It must be coupled with a continuous year-round output. Yet the non-steerable means of power generation do not offer such continuity and therefore do not constitute substitutes for incumbent nuclear electricity.

It follows from these considerations that there are presently no means of baseload generation – i.e., of supplying steerable power for more than 6,000 hours per year – which would allow an operator to compete with EDF on the supply market.

This brings us to the next question of whether it would be possible to duplicate the incumbent nuclear plants.

Here again, the answer is no. The non-reproducible character of the incumbent nuclear plants results from the following three factors.

The first factor is economic. The competitiveness of the nuclear plants is judged to be "unmatchable". This results mainly from the fact that the incumbent plants are largely amortized, and the efficiency gains arising from the serial building and operation of the nuclear units. The high performance of the French nuclear power plants is therefore not rooted solely in a technology – nuclear technology – but also in a specific historic context: the nuclear program of the 1970s.

The second factor is practical and relates to the virtual impossibility of finding an available physical site on which to build such an infrastructure.[16] The only sites on which the construction of new nuclear plants seems possible are those already occupied by such facilities. And those sites are owned by the incumbent operator.

The third factor is contextual. The French baseload market is sometimes described as being in an overcapacity situation, with no short or medium term need for the development of new capacities.[17] In this situation, it is unlikely that the price dynamics will, at least in the short term, motivate alternative suppliers to invest in new baseload generation capacities.

The Champsaur Commission's report on the evolution and regulation of electricity tariffs perfectly summed up this point by noting that duplication of the nuclear power fleet by alternative suppliers "is neither desirable nor feasible".[18]

Without any real substitute and any potential substitute, the essentiality of nuclear electricity for the activity of supplying electricity in the French market, therefore, appears to be established.

The financial strength of the essential facility owner's competitors

In its opinion of 21 January 2019, the ADLC points out that "the three main players on the retail electricity market, EDF, Engie and Total, are incumbent operators in the energy sector, vertically integrated upstream, with largely comparable turnover, technical resources and capacities to invest in production".[19]

This observation calls for two comments.

First, the emergence of integrated operators who follow the model of the incumbent operator is the homothety sought by the authorities when they opened the sector up to competition. However, due to the above-analyzed essentiality of the nuclear plants, that homothety has not been possible.

Second, the assessment of a facility's essentiality is never based on the financial solidity of the operator seeking access to it.[20] Rather, the assessment is conducted objectively and is based on an analysis of how a rational economic player would behave. All that matters is the existence of and access to economically comparable alternative resources.

What this economic analysis of nuclear power in France shows is the existing path dependency slows down the opening of the electricity market to new entrants and thus the development of alternative sources such as renewable energies. Yet nuclear power in France, although very firmly implemented and strongly supported politically and even socially (see Chapter 3),[21] is being questioned. Supporters of the nuclear industry recall that it emits very few GHG and call it a "clean energy". They also highlight the great success that it represents in France as it has endured for years, with security of supply, safe access and cheap prices for customers. Yet these arguments are now being discussed. Critics rose after the Fukushima accident but not as much as one might expect.[22] Presently, most of the criticism lies in the costs of the nuclear industry and the age of nuclear plants. Against this background, an important legislative step in 2015 led to the goal of reducing the share of nuclear power in electricity to 50% by 2025. However, this requires compensating with renewable sources, given that fossil energy does not constitute an alternative compatible with the GHG reduction target. As renewable energy generation grows fast but not fast enough, the 2015 target of nuclear power reduction has been postponed to 2035.

The promotion and slow rise of other energy sources and decentralized governance

The previous section highlighted that the French socio-technical system is tailored to promote the nuclear industry, which implies a centralized top-down governance scheme. It showed there are some strong arguments to maintain electricity generation as it is. In fact, the French authorities have frequently put off the planned closure of power plants[23] and postponed the age limit for their shutdown and dismantling.[24] However, a look at the French energy mix shows a slow rise in renewable energy generation, especially in the electricity mix. This comes along with the promotion by national authorities of a more decentralized and local governance.

The French energy mix

In France, 54.6 % of the consumed energy is produced domestically.[25] Since 1990, the consumption of coal and oil has decreased by 63% and 17%, respectively,

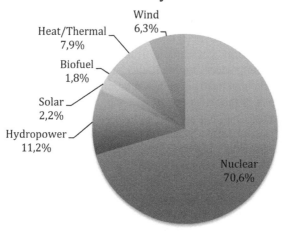

Electricity Generation Mix

Wind 6,3%

Heat/Thermal 7,9%

Biofuel 1,8%

Solar 2,2%

Hydropower 11,2%

Nuclear 70,6%

Figure 1.2 French electric generation mix in 2019.

while nuclear and natural gas increased by 28% and 46%, respectively. In the meantime, the consumption of renewable energy has almost doubled. Since 2005, it even increased by 70%, essentially thanks to heat pumps, wind and biofuels. In spite of these good trends, in 2019, only 11.6% of primary energy consumption came from renewable sources, more than 40.3% came from nuclear sources, about 15.2% came from natural gas, 29.1% from oil and 3% from coal. Within the renewable mix, solid biomass (essentially wood for heating) accounts for 4.2%, hydropower for 2%, biofuels for 1.4%, heat pumps for 1.1%, wind for 1.2% and others for 1.6% (Ministry of Ecological Transition, 2020).

As regards electricity in particular, Figure 1.2 shows that nuclear accounts for more than 70% of net generation in 2019, while renewables are behind with about 21% of the electricity produced. Hydropower is the biggest share among renewables but wind and solar have increased a lot with respect to 2018 (+21.2% and +7.8%, respectively).[26]

The French legal context promoting the role of new local actors in energy governance

In the energy sector, there have been two major legislative steps recently: the 2015 Law on Energy Transition for Green Growth (ETGG) and the Climate and Resilience Law of 2019. These pieces of legislation fix the national energy and climate targets and contain provisions on actors, tools and the organization of the energy sector. Moreover, based on the ETGG law, a Pluriannual Energy Program was adopted. It was revised in 2018 for the years 2019–2028. It determines the energy mix and sets an action program to meet the goals defined by law (see

Table 1.1). Another piece of legislation is now expected in 2023. This shows how quickly the sector and public policy are expected to evolve.

As seen earlier, the opening of energy markets, in particular electricity, is the outcome of a liberalization process initiated by the EU. This has enabled new operators to produce energy or sell energy services to final users. As renewable energies are a very context-based source of energy, and as they were promoted in EU law, it allowed local actors (local governments, citizens, individually or collectively) to participate more in depth to energy governance, as any other economic player.

However, this greater involvement of local governments is, in France, also the result of a general decentralization process (see Chapter 6), in which the central government has delegated energy-related missions to local authorities (Dreyfus and Allemand, 2018). Here, local governments are included as public authorities and political actors. Indeed, art. L 100-2 of the French Energy Code provides that the state acts "in coherence" with local governments and their groupings and also involves private companies and citizens. It means that local governments should support the implementation of national policy. This cooperation deals with, among others, energy demand management, promotion of energy efficiency and self-sufficiency; fighting fuel energy poverty; diversifying energy sources and in particular reducing the share of fossil fuel consumption and increasing renewable energy; fostering biomass use; and developing "positive energy territories" in areas where the local production of energy and its consumption is, as much as possible, locally met (see Chapter 9). The delegation of missions to local governments is also materialized by the numerous planning instruments used in different sectors (urban planning, transports, waste management). Most of these planning instruments specify environmental goals to be achieved locally through local public policies. In particular, municipalities and their groupings of over 20,000 inhabitants are bound to adopt local territorial climate-air-energy plans to tackle these three issues together. *Régions* have also enacted climate-air-energy regional schemes to translate the national objectives to the local context.

Another way to enhance local governance is through the development of self-consumption for individuals, private businesses and local public authorities. As regards self-consumption, the French government acknowledged it in law by an Ordinance of 2016[27] (that is even before the adoption of the European Clean Energy Package that strongly promotes it). It is defined as

> The self-consumption operation is collective when the supply of electricity is made between one or more producers and one or more final consumers linked together within a legal entity and whose extraction and injection points are located in the same building, including residential buildings. A collective self-consumption operation can be qualified as extended when the supply of electricity is carried out between one or more producers and one or more final consumers linked together within a legal entity whose extraction and injection points are located on the low-voltage network and meet the criteria,

particularly geographical proximity, set by decree of the minister in charge of energy, after consultation with the Energy Regulation Commission.

(Art. L 315-2 Energy Code)

It involves geographical proximity between the participants. However, at the end of 2019, only 16 collective self-consumption projects were working. Individual self-consumption is much more widespread with 52,096 individual plants in operating in 2019, representing 267.26 MW. There were 28,305 at the end of June 2018 and 13,877 one year earlier, showing a great increase.[28]

In 2019, the legislator transposed the legal category created by the EU (see above), that is the RECs, in art. L 211-3-2 Energy Code. It holds very similar features to the French definition of self-consumption: it is an autonomous legal body, with voluntary participation. Members or stakeholders of different kinds (local governments and their groupings, small commercial companies, individuals) located close to the production unit control it. Its goals are environmental, social and economic but they are not aimed at profit-making. It can produce, consume, stock and sell renewable energy, share it within the group of members/stakeholders and access all the relevant energy markets. However, its scope is wider and goes beyond the mere generation of electricity. It also implies working on energy demand. It remains to be seen whether a new legal category can foster the development of self-consumption projects.

Conclusion

The French electricity mix remains largely dominated by nuclear power despite an important increase in the share of renewable energy. This chapter showed that there is an important institutional and infrastructural path dependency. As a consequence, and as renewable energies are local resources,[29] it is only gradually that local actors – be they local authorities, individuals, or communities, or even small commercial companies – penetrate the sector. This development is largely determined by the various legislative steps, which facilitate the creation of new legal bodies and remove barriers in the market. This opening largely answers expectations from local actors (as the following chapters show) but also largely emanates from European law, which binds French authorities, and has triggered the liberalization of the energy market as well as the inclusion in the political agenda of climate change concerns and the need to promote renewable energy as a top priority. Surely yet, as long as the debate among French policymakers over the ecological footprint and non-substitutability of nuclear energy is not settled, a radical change toward more renewable energy, expected by some political actors and citizens, will not be achieved.

Notes

1 The first European project after World War II was the European Coal and Steel Community created by the Paris Treaty of 1951. Then, in 1957, EURATOM was created and promoted a single nuclear power market in Europe. Yet, within the general European institutional framework, it is only in 2007 in the Lisbon Treaty that an official

legal basis awarded the EU jurisdiction in the energy sector. Before that, the EU intervened based on its jurisdiction on competition and internal market matters. Then, when an environmental legal basis was introduced in a the Single European Act in 1997, it provided to the EU another ground for action in the energy sector. Now, since the 2007 Lisbon Treaty, a specific legal basis, art. 194, dedicated to energy has been adopted. It enables the EU to define its own "EU energy policy". This is a shared competence with the member states. The EU must therefore act in compliance with the principles of subsidiarity and proportionality.

2 This package includes four directives and four regulations. Energy Performance in Buildings Directive (EU) 2018/844: the Directive sets specific provisions for better and more energy-efficient buildings. It updates and amends many provisions from Directive 2010/31/EU. Renewable Energy Directive (EU) 2018/2001: the Directive sets a binding target of 32% for renewable energy sources (RES) in the EU's energy mix by 2030, with a possible review for an increase in 2023. It also includes provisions for mainstreaming RES in the transport and heating and cooling sectors. Energy Efficiency Directive (EU) 2018/2002: the Directive sets a target of 32.5% for energy efficiency for 2030, compared with a baseline scenario established in 2007, with a possible upward revision in 2023. It also includes provisions extending energy savings obligation and heat meters remote reading. Governance of the Energy Union Regulation (EU) 2018/1999: the Regulation sets a new governance system for the Energy Union. Each member state is to establish an integrated ten-year National Energy and Climate Plan (NECP) for 2021 to 2030, with a longer-term view toward 2050. The plan is to outline how the member state will achieve its respective targets. Electricity Regulation (EU) 2019/943 on the internal market for electricity: the Regulation sets principles for the internal EU electricity market. It focuses mainly on the wholesale market as well as network operation. In that regard, the Regulation includes provisions that affect certain articles in the electricity network codes and guidelines. It sets, for instance, a new bidding zone review process and establishes regional coordination centers, replacing the regional security coordinators, and complements the transmission system operators' roles on a regional scope. Electricity Directive (EU) 2019/944: the Directive sets rules for the generation, transmission, distribution, supply and storage of electricity. It also includes consumer empowerment and protection aspects. In addition, the market design Directive sets provisions for distribution system operators' flexibility procurement. Risk Preparedness Regulation (EU) 2019/941: the Regulation requires member states to prepare plans on how to deal with potential future electricity crises. They are to use common methods and identify the possible electricity crisis scenarios, at both national and regional levels. Risk preparedness plans shall be based on these scenarios. ACER Regulation (EU) 2019/942: the Regulation updates the role and functioning of the European Union Agency for the Cooperation of Energy Regulators (ACER). The Clean Energy Package also increases the competence of the ACER in cross-border cooperation. Moreover, it adapts the ACER's tasks to the new regulatory framework established by the Clean Energy Package, such as for the decision on the system operation regions and the monitoring of regional coordination centers.

3 There were three waves of liberalization: the first "energy package" in 1996 for electricity and 1998 for gas, the second "energy package" in 2003 and the third "energy package" in 2009.

4 Directive (EU) 2019/944 of the European Parliament and of the Council of 5 June 2019 on common rules for the internal market for electricity and amending Directive 2012/27/EU.

5 Directive (EU) 2018/2001 of the European Parliament and of the Council of 11 December 2018 on the promotion of the use of energy from renewable sources.

6 Th. Dahan, Hearing in connection with the Commission of inquiry on electricity prices, French National Assembly (2014).

7 European Commission Decision of 12 June 2012, SA.21918, point 156.
8 Ibid., §155.
9 On the concept of "workable competition", see ECJ, 25 October 1977, *Metro SABA*, 26–76, point 20.
10 Report of the commission on the organization of the electricity market, 2009 (Champsaur Report).
11 CFI, 12 June 1997, *Tiercé Ladbroke SA v. Commission*, T-504/93, Rec. p. II-923, spec. point 131.
12 Unofficial translation, NOME bill.
13 Report of the commission on the organization of the electricity market, 2009 (Champsaur Report), p. 27.
14 ADLC, opinion no. 19-A-01, 21 January 2019, supra, point 209.
15 We refer here mainly to wind and solar power.
16 ADLC, opinion no. 10-A-08, 17 May 2010, supra, point 19.
17 See, in particular: P. Champsaur, Hearing in connection with the Commission of inquiry on the real cost of electricity, French Senate (2012). This assessment can be understood only if one compares the total output of the baseload production means installed in France with the total consumption of baseload electricity.
18 Cf. footnote 13, p. 11.
19 ADLC, opinion no. 19-A-01, 21 January 2019, supra, point 213.
20 See in this respect the *Oscar Bronner* decision (ECJ, 26 November 1998, *Oscar Bronner v. Mediaprint*, C-7/97, Rec. pp. I–779).
21 Some authors have investigated how the acceptability of the nuclear industry was "constructed" by political institutions (Topçu S. (2013) *La France nucléaire. L'art de gouverner une technologie contestée*, Paris: Seuil).
22 In comparison, Germany decided to shut down its 17 nuclear power plants immediately after 11 March 2011 and by 2022.
23 For example, French President Francois Hollande, in 2012, in the aftermath of the Fukushima accident, promised to shut down within a year the Fessenheim plant. Yet it is only in 2021 that the first reactor was turned off.
24 In 2021, the French Nuclear Safety Agency authorized some of the oldest reactors, which were meant to function for 40 years, to have their lifespan expanded to 50 years.
25 This is based on the fact the nuclear power and the heat it produces are considered by IEA as a primary energy source. However, nuclear energy is indeed the result of a transformation of fuel such as uranium that is entirely imported (which therefore tempers how the level of energy autonomy should be appreciated).
26 https://bilan-electrique-2019.rte-france.com/production-totale/#.
27 Ordonnance 27 July 2016, based on ETGG, (ratified in 2017).
28 At the same time, Germany accounts for 1.5 million self-consumers and the UK 750,000. The answer of the Ministry of Ecologic Transition and Solidarity to the question of Senator Jean-Marie Bocquel, 2020 (www.senat.fr/questions/base/2018/qSEQ181208040.html).
29 This is not completely the case and hydropower often works with mega-structures owned and managed by private or public national actors. In France, EDF is the main incumbent in this sector.

References

Dreyfus, M. and Allemand, R. (2018) 'Three years after the French energy transition for green growth law: Has energy transition actually started at the local level?', *Journal of Environmental Law*, 1, 109–133. https://doi.org/10.1093/jel/eqx031

Heldeweg, M.A. and Saintier, S. (2020), 'Renewable energy communities as 'socio-legal institutions': A normative frame for energy decentralization?', *Renewable and Sustainable Energy Reviews*, vol. 119. https://doi.org/10.1016/j.rser.2019.109518

Lowitzsch, J., Hoicka, C.E., and van Tulder, F.J. (2021) 'Renewable energy communities under the 2019 European Clean Energy Package – Governance model for the energy clusters of the future?', *Renewable and Sustainable Energy Reviews*, 122. https://doi.org/10.1016/j.rser.2019.109489

Ministère de la transition écologique (2020), *Chiffres clés de l'énergie 2020*, https://www.statistiques.developpement-durable.gouv.fr/chiffres-cles-de-lenergie-edition-2020-0

2 Energy transition in Japan

The political landscape and civil society's contribution

Masayoshi Iyoda

The IPCC Special Report on 1.5°C of global warming highlights that global society needs to achieve net-zero carbon by 2050 to meet the Paris Agreement 1.5°C goal. Consequently, governments will have to reduce carbon pollution to "net-zero" by ending fossil fuel use completely.[1] As stated in the IPCC 5th assessment report, the nuclear power option, on one side, is problematic because of its cost, waste concerns, lack of public support and risk of uncontrollable accidents (IPCC, 2014). The Fukushima Daiichi nuclear disaster in 2011 reinstates the concerns. Climate risks and nuclear ramifications have made global civil society support a long-term vision of a 100% renewable energy society while achieving the highest energy-saving effort (CAN, 2014).

There is, however, a need to understand how multi-level governance, including civil society, is constructed and functions to understand how to realize a fair and just energy transition to a 100% renewable and climate-safe society. This is particularly important in Japan where, in contrast to the progressive global recognition, a 100% renewable future has been perceived as "an unrealistic aspiration", rather than a serious political imperative, by the government and the general public.

This chapter describes the political landscape and the development of climate and renewable energy advocacy, illustrating an overview of the 100% renewable energy proposals and initiatives addressed by civil organizations in Japan. Civil society and the associated organizations have been defined by the United Nations as

> non-state, not-for-profit, voluntary entities formed by people in the social sphere that are separate from the State and the market, which represent a wide range of interests and ties. They can include community-based organizations as well as non-governmental organizations (NGOs).

The chapter specifically discusses the motives and barriers to seeking a 100% renewable society, laying down the background to understand the need for multi-level governance for such a renewable transition.[2] Energy transition aims at decarbonizing the entire energy system, including the electricity sector. In this chapter, the 100% renewable society refers to the holistic aggregation of the relevant visions, plans and actions to achieve such a system.

DOI: 10.4324/9781003025962-4

Governmental preferences and their political background

In Japan, about 85% of greenhouse gas emissions come from fossil energy, thus climate policy inevitably corresponds with energy policy. Energy policy is under the responsibility of the Ministry of Economy, Trade and Industry Japan (METI) and its extra-ministerial bureau, the Agency of Natural Resources and Energy (ANRE). Both have developed a strong connection with industry groups, including the Japan Business Federation ("Keidanren") and the Federation of Electric Power Companies of Japan (FEPC, or "Denjiren"). Their priorities have been energy security and economic efficiency, rather than environmental conservation, including climate change mitigation.

On the other hand, the Ministry of the Environment (MOE) is, in principle, the governmental authority for implementing government policies and regulations for adapting and mitigating the negative impacts of climate change. It is given, however, a relatively weak status among other government agencies and it does not have much power to intervene in energy policy stipulated by METI and ANRE under the Basic Act on Energy Policy. The presence of these ministries seems to be unbalanced, where the general account budget allocated to METI was about 1,200 billion JPY, more than double the approximately 450 billion JPY of MOE in 2019.[3] Also, from the number of staff perspective, the METI and ANRE staff counted to around 5,000, while approximately 2,000 staff members were with the MOE in 2019.[4] These discrepancies between METI and MOE have been reflected in the overall Japanese climate and energy policy.

One illustrative example of conflict between METI and MOE has been whether and to what extent coal power generation shall be continued and expanded in Japan. Coal power is obviously the largest source of CO_2 emissions that should be eliminated for a decarbonized future and successive environmental ministers have warned about the issue. However, the MOE has not been able to override the METI's decision to allow utilities to build new coal power plants[5] (Kiko Network, 2018). With limited power, MOE developed a tendency to focus only on issues (often of trivial nature) where it can exercise influence, for example, raising environmental awareness on energy saving in households.

As indicated by the strong position of METI, the Japanese government has not been giving a high priority to climate change mitigation nor renewable energy promotion. Rather, nuclear power has been considered an integral part of Japan's energy supply system, and its benefits arguably have been emphasized: it has contributed to energy diversification, reduced dependence on oil, was produced at a stable price, and has been emissions-free in the generation stage (Lesbirel 2004; Vivoda 2014). All of Japan's electric utility monopolies, with the exception of Okinawa Electric Power Company, own nuclear power plants and, prior to the Fukushima nuclear disaster, nuclear power was a key part of their electricity supply portfolios. The rising energy costs due to the nuclear shutdown and increased cost of fossil fuel imports have hurt utilities' profitability, especially as they have been unable to increase service prices (Vivoda and Graetz, 2015).

The traumatic "Oil Crisis" experience in the 1960s compelled the Japanese government to prioritize energy security through fossil and nuclear development. Its energy policy has also been largely affected by high energy-intensive heavy industry sectors, including iron/steel, automotive, cement, electrical machinery and oil/petrochemicals production (InfluenceMap, 2020). The myriad of state-industrial allies has become a political barrier, as they actively and strategically engage in energy policy processes where they often oppose potentially effective climate policy. It is furthermore remarkable that the Japanese government frequently emphasizes the downsides of renewable energy, for example, in public school textbooks, leaving the public with the impression that renewable energy is unstable and costly.

Changes within the national government

A small but apparent change, nevertheless, is emerging within the government, reflecting the ever-increasing international pressure to focus more on renewable energy development. As an example, though METI is yet reluctant, some other ministries are taking up the 100% renewable target in the institutional agenda. So far, the Ministry of the Foreign Affairs (MOFA), the Ministry of the Environment (MOE) and the Ministry of Defense (MOD) have committed to achieving the 100% renewable target.

There is, however, a twist of relations among these ministerial authorities. The interesting fact is that the first pledge of 100% renewables within the government was made by MOFA in May 2018, not MOE, which should have prioritized environmental protection.[6] This unusual sequence can be linked to the personal leadership of Mr. Taro Kono, who took the ministerial position at MOFA, while the political leadership of the Minister of the Environment was weaker. In May 2020, Mr. Kono became Minister of Defense and decided that the MOD would become the third ministry to officially announce its intention to institutionally achieve the 100% renewables target. Mr. Kono, as a cabinet member, kept his silence regarding his own opinion on the nuclear issue in order to avoid any political conflict within the cabinet, but it is well known that he has supported nuclear phase-out in the past, which is exceptional among the ruling party members.

Meanwhile, the government reviewed its energy policy from 2017 to 2018. In July 2018, after an intensive discussion, especially on the controversial position of nuclear power, the government formulated its 5th Basic Energy Plan, a fundamental framework of the energy policy. The 5th Basic Energy Plan still emphasized coal and nuclear as "important baseload sources of electricity", prioritizing thermal and nuclear powers rather than renewable energy in the overall Japanese electricity portfolio. It, however, identified renewable energy as "a main source of electricity" for the first time ever in Japanese political documents on energy issues, and most of the climate policy specialists welcomed that the "main source of electricity" status was finally bestowed on renewable energy.

In October 2020, one month later than China's lead, Prime Minister Yoshihide Suga announced that the Japanese government had set a goal to achieve carbon neutrality, in other words, net-zero emissions by 2050. He expressed that he would promote renewable energy with maximum commitment but also continue to use nuclear power. Experts see Japan's move to embracing the climate target of carbon neutrality by 2050 as opening the way for the struggling nuclear industry to fire up again, nearly a decade after the Fukushima disaster shut down most of the country's nuclear reactors (World Economic Forum, 2020). A majority of people, however, have opposed the pro-nuclear policy and have supported nuclear phasing out since the Fukushima disaster, making the government hesitant of allowing electric utilities to build new nuclear power plants.

A brief history of civil society's advocacy on climate and energy issues

These changes at the national government level are partly the reflection of increasing civil pressure, though the history and magnitude of climate NGOs in Japan are relatively limited with respect to those of their counterparts in Europe and the United States. Until the early 1990s, the Japanese government and the public alike were relatively less conscious of the global warming issue. The remarkable turning point was when Japan hosted the third Conference of Parties (COP3) of the United Nations Framework Convention on Climate Change (UNFCCC) in 1997. It worked as a "wake-up call" for the Japanese people and civil society to understand the significance of climate change. Just before COP3 was held in Kyoto, Kiko Forum,[7] a concerned citizens' network, was established to enact sufficient pressure onto governments to develop a new and effective climate protocol. The activities of Kiko Forum were initially led by Japanese environmental NGOs, such as Citizens' Alliance for Saving the Atmosphere and the Earth (CASA), Shimin Forum 2001 and the Japanese branches of the influential international environmental NGOs (such as Greenpeace and WWF). Kiko Forum facilitated national civil movements through demonstrations, a series of quality seminars and symposiums, petitions, advocacy and lobbying at COP3. Their enthusiastic and professional activities played an essential role in addressing public climate concerns and were well recognized by the mass media in Japan.

In 1998, Kiko Network, which succeeded Kiko Forum, started its activities to urge the Japanese government to ratify and implement the Kyoto Protocol. In 2001, however, the then Bush administration announced the United States' denial to participate in the Kyoto Protocol. This constituted a significant threat to the global climate regime under the Kyoto Protocol and the UNFCCC. Japanese diplomacy has traditionally given the highest priority to its relationship with the United States, therefore, Japan faced a difficult choice as to whether it should follow the US trail or take an "independent" path. Japanese industry groups, in addition, insisted that the Kyoto Protocol was an unequal treaty and pressured the government to stop Japan's ratification of the Kyoto Protocol. Eventually, with the Diet's approval, the Japanese government decided to ratify the Kyoto Protocol in 2002, after receiving massive demands from civil society organizations, including Kiko

Network, to do so. After the ratification by the Russian Federation, the Kyoto Protocol eventually entered into force in February 2005.

After the Kyoto regime started being implemented, the climate NGOs began lobbying the Japanese government to set more ambitious emission reduction targets and introduce effective climate policies and measures, including a carbon tax, a cap and trade scheme and a feed-in tariff (FIT) for renewable energy diffusion. The Liberal Democratic Party (LDP), *Jiminto* (almost continuously in power since its foundation in 1955, only with the exception of a period between 1993 and 1994, and again from 2009 to 2012), has been not responsive to civil society organizations' advocacy. Also, the Keidanren, the organization of heavy industry and energy companies, was exercising a strong influence on the national climate and energy policy. They feared the potential economic disadvantage caused by tighter climate actions.

In 2009, the Democratic Party of Japan (DPJ), *Minshuto*, came to power in Japan with a manifesto including setting a climate target to reduce greenhouse gas emissions by 25% by 2020 against the 1990 level. This manifesto to set a higher climate target and introduce a carbon tax, cap and trade and FIT, was influenced by a campaign called "MAKE the RULE", which civil society organizations coordinated. This marked a case where civil organizations exercised their influence upon the government agenda, though unfortunately, the DPJ was not fully able to pass the associated legislations when in power.

After the Great East Japan Earthquake caused the Fukushima nuclear disaster in March 2011, the nuclear aspect of energy policy became a top priority for the political agenda of the government and society in Japan. After the Fukushima disaster, the massive wave of anti-nuclear protests provoked tensions between anti-nuclear protesters and climate-oriented organizations (Cassegard, 2017). Even a few but bold protesters denied climate change science because they perceived it as an excuse or a "lie" to expand and keep nuclear power by the government and electricity companies. Also, under the emergency situation right after the disaster, a number of people prioritized nuclear phase-out over climate protection and even environmentally minded people tended to have an impression that carbon pollution was less harmful than the risk of nuclear.

The climate aspect of the energy policy was rather overlooked in the post-Fukushima discourse. Only a small number of NGOs tried to continue providing systemic advocacy to achieve both a climate-safe and nuclear-free world. The renewable energy policy gained momentum when Prime Minister Naoto Kan of DPJ made the Diet pass a bill for FIT in exchange for his resignation in 2012. Since then, renewables have rapidly expanded in Japan, as Chapter 4 of this volume elaborates in detail. Simultaneously, 50 coal-fired power plant projects, however, have been newly planned since 2012 in Japan (Kiko Network 2018). Such a backward move in terms of climate protection was justified as a means to match up the electricity deficit caused by the shutdown of nuclear power plants after the Fukushima disaster.

In 2013, the DPJ government tried to reconsider energy policy through national debate and developed the "Innovative Energy and Environment Strategy", which seeks nuclear phase-out by the 2030s. However, after the election at the end of

2013, the DPJ was defeated, and the LDP, led by Mr. Shinzo Abe, took office. Prime Minister Abe (at the time) expressed his intention to reverse the energy policy of the former government. His government developed the 4th Basic Energy Plan in 2014, which regarded nuclear and coal-fired power as "important baseload electricity", abandoning the "Innovative Energy and Environmental Strategy" and the other national debate outcomes the DPJ government stipulated. Under the Abe government, it became more difficult for climate NGOs to influence climate and energy policy at the national level because political conservatism, typical to the Abe regime, prevailed.

Although having a certain policy impact as shown earlier, civil society's influence is limited in Japan and much weaker than that of the Western countries. Schreurs (2002) points out that because environmental NGOs in Japan do not consolidate their influence sufficiently, MOE has not received the same support as its counterparts in the West. Thus, MOE had to independently promote environmental policy and seek support from abroad, a few progressive domestic organizations and politicians. Hasegawa and Shinada (2016) argue that Japanese politics and its policy-making process have been inflexible and that Japan's climate change policy is predominated by each ministry, governmental committees and industry groups, but not civil society such as NGOs. Furthermore, media and journalists highly depend on press releases given by ministry and industry groups, rather than carrying out their independent research. There is limited news coverage on the issues that civil society advocates. Furthermore, this media bias has been exasperated in recent years, compared with 1997 when a constructive media-public relation during and after COP3 in Kyoto raised momentum.

At the international level, it is expected that the US Biden administration will aim to achieve 100% clean energy and bring new geopolitical dynamics to global climate politics. Japan cannot avoid pressure from the United States, so Japan may consider climate and energy policy more seriously.

From an economic aspect, renewables' cost was more expensive in the 1990s, 2000s and 2010s, than in the 2020s. Therefore, before, most policymakers did not believe that renewables could be the primary source of electricity, though there were rare examples of proposals to be realized, as in the FIT case mentioned previously. A study suggests, however, that the cost of renewable electricity in Japan is getting more competitive (Carbon Tracker, 2019). Thus, the primary choice should shift from the current debate of "to choose renewables or not renewables" to the next dimension of "to choose renewables or much more renewables", suggesting the irreversible dissemination of renewable electricity.

Civil society's actions to promote energy transition to renewables

Though relatively weak, there have been various types of actions for civil society to promote renewable energy expansion in Japan. These activities have created an enabling environment for addressing energy transition in Japan. At the strategic level, civil society organizations have made policy proposals and committed to advocacy activities both at international and domestic political arenas related to climate and energy issues, including successive UNFCCC Conference of Parties and the Group of Seven (G7), the Group of Eight (G8) and the Group of 20 (G20). Domestically, civil society organizations have tried to make their voice heard at the National Diet and governmental committees.

Various actors, such as business leaders, local governments, universities and civil society organizations, have conceived a 100% renewable energy vision. The most famous initiative of the 100% renewables pledge in Japan is "RE100", a global initiative from the most influential businesses aiming to achieve 100% renewable electricity company-wide by 2050. Coordinated by the Climate Group (an international non-profit founded in 2003, with offices in London, New York and New Delhi) and CDP (a not-for-profit charity that runs the global disclosure system for investors, companies, cities, states and regions), RE100 was launched in 2014 at Climate Week NYC and was initially joined by companies based in Europe and North America. Climate NGOs have encouraged Japanese companies to join RE100, but it took almost three years before Japan-CLP, a group of progressive businesses on the climate issue, became in charge of the regional partner of RE100, and RICOH became the first Japanese company to join the RE100 in 2017. RICOH is well known as a leading company in the printer and copier market and one of the progressive business actors in the field of climate action. After RICOH broke the silence, more than 58 Japanese companies (including Sekisui House, ASKUL, Daiwa House and AEON) joined this global initiative and pledged to achieve a 100% renewable energy. Currently, Japan is one of the top RE100 countries in terms of the number of member companies.

In 2017, Climate Action Network Japan (CAN-Japan), a network of Japanese climate NGOs, launched "100% Renewable Energy Platform Japan", supported by the Global 100% Renewable Energy Platform. As RE100 is open to corporate giants, the platform intended to take over pledges made by relatively small-scaled actors from business sectors, local governments and education institutions. The platform registered 16 actors pledging 100% renewable targets, such as Fukushima Prefecture and Chiba University of Commerce (CUC), which is the first university to declare a 100% renewable target in Japan. Also, the platform collects 110 approbations and shows the expanded demands for a 100% renewable society. In 2019, the RE Action, a brand-new initiative seeking 100% renewables, was launched by Green Purchasing Network (GPN) and other supporters, including Japan-CLP. RE Action is also a pledging campaign of 100% renewables in Japan targeted at smaller businesses, local governments and civil society organizations. Ninety-eight actors have joined RE Action, including 69 companies, five local governments/governmental institutions, 11 non-profit organizations and counting.[8] After a conversation between RE Action and 100% Renewable Energy Platform Japan, the latter stopped collecting 100% renewable pledges, and the former took over the activity to collect 100% renewables pledges. With increasing pledges and approbations, the initiatives, platform and campaign made a step forward in disseminating the vision of 100% renewables in Japan.

In parallel, several policy proposals and roadmaps of a 100% renewable energy society targeted for Japan have been developed. WWF Japan, for example, developed a 100% renewable energy scenario, which assumes that in 2050, all of Japan's energy will be supplied by renewable energy sources while final energy consumption could be reduced to about half (WWF Japan, 2017). The rest of the energy demands will be met by photovoltaic, wind, hydro, biomass, geothermal and tidal/ocean power. These proposals emphasize multiple benefits of a transition to a 100% renewable energy society. In addition to the apparent value of protecting human health by improving air quality, renewable energy has been identified as a means to secure job opportunities and contribute to reducing the monetary outflows associated with fossil fuel imports. It means that the transition to 100% renewables has environmental, social and economic sense. Also, these benefits strongly link with Sustainable Development Goals (SDGs), especially SDG7, which is to ensure access to affordable, reliable, sustainable and modern energy for all, and SDG11, which is to make cities and human settlements inclusive, safe, resilient and sustainable.

Apart from the strategic policy proposal, civil society organizations take up a bottom-up approach, with one notable example being "Citizens' Co-owned Renewable Energy Power Plants (CCREPP)". CCREPP is a community-based project to build and operate a renewable energy power plant coordinated by civil society members, with the funds mainly provided by citizens' donations. The first attempt to build a CCREPP was in Miyazaki Prefecture in 1993, followed by a project in Shiga Prefecture in 1998, then CCREPP activities started to spread nationwide (Toyota, 2016). By 2017, it grew to approximately 1,030 plants, which is equivalent to 90 MW. Although the total capacity is much smaller than a large-scale nuclear or coal-fired power plant, these activities attracted people and mainstreamed renewable energy in people's minds. The expansion of CCREPP showed the possibility and expectation of a small-scale distributed system of renewables that is much less harmful than conventional energy sources. Each year, leaders and supporters of CCREPP have organized a nationwide conference ("National Forum for CCREPP"), which functions as a cradle to facilitate civil society's renewables movement in Japan. In 2019, the 11th National Forum for CCREPP adopted an appeal for participants to achieve a 100% renewable society in Japan by 2050 at the latest, which indicates that bottom-up and grassroot CCREPP activities are connected with the national policy debate on energy transition.

In addition, local governments have their own initiatives to adopt ordinances to increase renewables, which are mainly to seek climate protection and renewable diffusion, but also to make sure the benefits arising from such developments are in the hands of their residents. Some of these ordinances clearly recognize renewable resources in their area, such as sunlight and wind, and state that they should belong to and be used for the local community. For example, "Iida City Ordinance for Building Sustainable City through Introduction of Renewable Energy" formulates policies and measures to support renewable projects led by the local community. It is also remarkable that "Kyoto City Ordinance of Global Warming Counter Measures" set a legal obligation for new large-scaled buildings to install renewable energy facilities, even in the absence of such a policy at the national government level. Civil society has supported discussions and the adoption of city ordinances as an essential part of local energy governance.

After the liberalization of electricity retailing in 2016, civil society organizations decided to launch a "Power Shift" campaign. This campaign encourages people, offices and local governments to switch from a conventional electricity company, whose portfolio is largely dominated by coal and nuclear, to a new electricity retailing company (Power Producer and Supplier, PPSs) that has the strong motivation to promote sustainable renewable electricity and avoid coal and nuclear power. This campaign, coordinated by FoE Japan, is to accelerate energy transition by raising awareness of individuals and companies on energy issues through sharing information on electricity system reform and new opportunities the reform may bring. The Power Shift (2019) asked a number of people about preferring environmentally friendly electricity to nuclear power but they needed appropriate information and certain triggers for switching. Switching from conventional electric power companies to PPS has risen to approximately 14% in electricity sold in the low-voltage sector, such as households, as of December 2018. These actions from civil society can accelerate energy transition and are an essential part of local energy governance.

Conclusion

As discussed previously, the government, under pressure from industry groups, has demonstrated limited support for renewable energy and does not seem to be fully ready to engage in a drastic energy transition. Civil society and progressive actors face multiple barriers before promoting the 100% renewable vision to its full extent. The lack of political will and the relevant policy framework to expand renewables are enormous barriers. The LDP government admits that renewable energy use should be increased, but it is uncertain to what extent the government can exclude influence by the vested interests of the nuclear and coal industries. The government is, for example, introducing a capacity market as part of reforms to the electric power system. The capacity market is generally a mechanism to achieve long-term security of supply by offering payments to power generators for being available to generate at certain times and to demand response providers for being able to reduce electricity demand. It may function, however, as a de facto subsidy for nuclear and coal-fired electricity generating plants. With the market design as it is currently being considered, the size of the capacity market will likely be around 1.4 trillion yen, which will finally be levied from electricity bills (Matsukubo, 2019).

Nevertheless, the idea of a need to trigger energy transition to new energy sources has been expanding in Japan with a massive push by civil society organizations and progressive non-state actors. There are considerations about the various motives to seek 100% renewables. Firstly, there is no doubt that concern about the climate crisis is one of the primary motives for promoting renewables in Japan. Awareness has grown since the Kyoto Protocol and made people support sustainable and fair renewables.[9] Also, members and supporters of the 100% renewables initiatives, such as RE100 and RE Action, clearly explain their intention to address climate change. Secondly, nuclear phase-out is a significant motive, especially for the post-Fukushima era. It is observed that anti-nuclear movement groups, as well as climate-oriented organizations, clearly state that the option of 100% renewables should be a solution. Thirdly, promoting a vision of 100% renewables contributes to building a better reputation and a brand image in society, especially for businesses or universities. Some reports state that pledging a 100% renewable target attracts media journalists and has a large publicity effect. Fourthly, pressure in a supply chain becomes an essential factor for some companies. There is a sense of urgency among corporate institutions that they may be excluded from a global supply chain if they fail to show their commitment to energy transition. For example, Apple Inc., one of the members of RE100, has demanded that suppliers switch to 100% renewables to achieve 100% renewables as an entire supply chain. According to a press release published by Apple, some Japanese companies, such as Ibiden and Nidec, have committed to running their Apple production on 100% clean energy.

Civil society advocacy and movements will surely be persistent and continue to encourage a sustainable, fair and just transition to 100% renewables in Japan. In fact, media coverage regarding 100% renewables has been increasing in recent years (Figure 2.1). Among youth climate activists, a vision of 100% renewables is one of the key messages. At a climate march for Fridays For Future Kyoto, as part of the global youth climate movement started by Greta Thunberg, there have been messages seeking 100% renewables.

On the other hand, though 72% of Japanese people support an increase in renewable energy,[10] greater care may be necessary in Japan, where the general public has been accustomed to skepticism on renewable energy, adopting humble

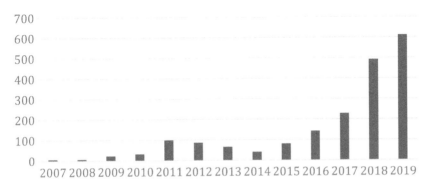

Figure 2.1 Media coverage including words of "100% renewable energy" (Searched in a news article database of Nikkei Telecon 21 by counting the amount of media coverage that included words which mean 100% renewable energy each year.)

incrementalism rather than strategic leaps based on understanding the climate imperative. In Japan, it has been observed that people show mixed attitudes toward renewables and prioritization. This may be the result of the repetitive discourse typical to the national government that renewables are too costly and unstable to expect progress. When expectations and ambition are limited, it is hard to build momentum to vitalize renewable expansion policies.

There are, also, a few controversial renewable projects raising concerns in local communities, especially between business operators and local residents, on issues such as the impact of photovoltaic facilities or wind turbine developments on nature and landscape conservation. Also, some have concerns about the kind of exploitation that big companies based in an urban area like Tokyo intend to make in the countryside, particularly when they build large-scale renewable energy stations. Local people fear that such developers would deprive local wealth without enough communication, ignore the community's consent or opposition, and impede the circulation of profits back to the local level. Through the environmental impact assessment process and other policies, such conflicts should be avoided. It is important for local people to lead or at least participate in local renewables projects to secure environmental integrity and the circular economy on the ground, partnering with various stakeholders such as experts and local government. This is exactly why local governance is needed to consolidate multiple aspects associated with further energy transition, as this entire volume discusses.

Notes

1 Unless CCS and CCUS technologies are successfully innovated and operated in a cheap, safe and harmless manner, there is no prospect of successfully developing and deploying CCS and CCUS promptly for the 1.5°C trajectory.
2 When we discuss "100% renewable energy", attention should be paid to sector coverage. Three sectors are often discussed: electricity, heat and transport. In many cases, 100% renewable actually means 100% renewable electricity.
3 Ministry of Finance (2019) "general account budget requests".
   https://www.mof.go.jp/budget/budger_workflow/budget/fy2020/sy010905.pdf, accessed 29 November 2020.
4 National Personnel Authority (Jinjiin) (2019) "Annual Report FY2019".
   https://www.jinji.go.jp/hakusho/pdf/sankou.pdf, accessed 29 November 2020.
5 Newly built thermal power plants combusting coal and gas may be a barrier to renewable energy expansion. Once a large-scale conventional power plant is built, it is expected to be in operation for decades and expel renewable energy diffusion. It has already occurred by Kyushu Electric Power Company after they refused to be connected and supplied from renewable energy power stations, such as solar photovoltaic, while the company operated nuclear power plants.
6 To elaborate, in May 2018, Mr. Taro Kono, the Minister of Foreign Affairs (at the time), announced that the MOFA, including diplomatic missions abroad, would seek to achieve 100% renewables and try to join the RE100. In the immediate reaction to the announcement, Mr. Masaharu Nakagawa, the Minister of the Environment (at the time), answered a question from a journalist at the press conference that "I assume that it is practically difficult only for the MOE join the RE100. As MOE, our responsibility is to consider how to promote carbon-free electricity use within the government, and to take the lead". It is a typical Japanese political statement suggesting the hesitance of

MOE to be in line with MOFA. In fact, he denied setting a 100% RE goal for MOE as an institution. A month later, however, Mr. Nakagawa eventually showed the intention to achieve 100% renewables at MOE, though as a long-term vision, following MOFA.

7 Japanese civil society formed the "Kiko Forum", which learned from the experience of the "Klima Forum" that successfully contributed to an agreement at COP1 in Berlin, Germany. "Kiko" means climate in Japanese.

8 REaction (2020) "Annual Report 2020". https://saiene.jp/news/1826.

9 Even in the 1990s, the cost of a solar photovoltaic module was too expensive to pay. Nevertheless, many people tried to install solar panels on their house's rooftop to "stop global warming", knowing it could produce deficits.

10 Mainichi Shimbun (2020). https://mainichi.jp/articles/20201107/k00/00m/010/139000c.

References

CAN (2014) 'CAN Position: Long term global goal for 2050', http://www.climatenetwork .org/publication/can-position-long-term-global-goals-2050-june-2014, accessed 31 August 2020.

Carbon Tracker (2019) 'Land of the rising sun and offshore wind: The financial risks and economic viability of coal power in Japan', https://carbontracker.org/reports/land-of -the-rising-sun/

Cassegard, C. (2017) 'Between government and grassroots; Challenges to institutionalization in the Japanese environmental movement' in C. Cassegard (eds) *Climate action in a globalizing world: Comparative perspectives on environmental movements in the global north*. New York: Routledge, pp. 149–170.

Day, T., Höhne, N., and Gonzales, S. (2015) 'Assessing the missed benefits of countries' national contributions Quantifying potential co-benefits', https://newclimateinstitute .files.wordpress.com/2015/10/cobenefits-of-indcs-october-2015.pdf

Hasegawa, K. and Shinada, T. (2016) 'Sociology of climate change policy – Can Japan change? (*Kikouhendouseisakunoshakaigaku – Nihonhakawarerunoka?*)', Showado, Kyoto.

InfluenceMap (2020) 'Japanese industry groups and climate policy', https://influencemap .org/presentation/Japanese-Industry-Groups-and-Climate-Policy-899704d005cb963 59cc5b5e2a9b18a84, accessed 30 August 2020.

IPCC (2014) *Climate Change 2014: Mitigation of climate change. Contribution of working group III to the fifth assessment report of the intergovernmental panel on climate change*. Cambridge, United Kingdom and New York, NY, USA: Cambridge University Press.

IPCC (2018) *Global Warming of 1.5°C. An IPCC special report on the impacts of global warming of 1.5°C above pre-industrial levels and related global greenhouse gas emission pathways, in the context of strengthening the global response to the threat of climate change, sustainable development, and efforts to eradicate poverty*. Geneva, Switzerland: World Meteorological Organization.

Jacobson, M.Z. et al. (2017) '100% clean and renewable wind, water, and sunlight all-sector energy roadmaps for 139 countries of the world', *Joule*, 1, pp. 108–121.

Kiko Network (2018) 'Japan coal phase-out: The path to phase-out by 2030', https://www .kikonet.org/wp/wp-content/uploads/2018/11/Report_Japan-Coal-Phase-Out_EG.pdf

Lesbirel, S. (2004) 'Diversification and energy security risks: The Japanese case', *Japanese Journal of Political Science*, 5(1), pp. 1–22. https://doi.org/10.1017/ S146810990400129X.

Matsukubo, H. (2019) 'The capacity market: An overview and issues', Citizens' Nuclear Information Center, https://cnic.jp/english/?p=4435.

REN21 (2017) *Renewables global futures report: Great debates towards 100% renewable energy*, https://www.ren21.net/wp-content/uploads/2019/06/GFR-Full -Report-2017_webversion_3.pdf.

RE100 (2020) 'RE100; Companies', http://www.there100.org/companies/, accessed 30 August 2020.

Schreurs, M.A. (2002) *Environmental politics in Japan, Germany, and the United States*. Cambridge: The Syndicate of the Press of the University of Cambridge. This Japanese edition, *Chikyukankyomondai no hikakuseijigaku*, published 2007 by Iwanami Shoten Publishers, Tokyo.

Toyota, Y. (2016) *Trends and developments of citizens' co-owned renewable energy power plants*, Sustainability Study 6. Tokyo: Hosei University, pp. 87–100. [In Japanese.]

Vivoda, V. and Graetz, G. (2015) 'Nuclear policy and regulation in Japan after Fukushima: Navigating the crisis', *Journal of Contemporary Asia*, 45(3), pp. 490–509, https://doi .org/10.1080/00472336.2014.981283.

Vivoda, V. (2014) *Energy security in Japan: Challenges after Fukushima*. Aldershot: Ashgate.

World Economic Forum (2020) *Japan wants to be carbon neutral by 2050. Here's what it will mean for nuclear energy*, https://www.weforum.org/agenda/2020/11/japan-climate -change-carbon-neutral-nuclear-power-fukushima-energy/.

Part II

Local government powers in the energy sector

3   Local authorities and energy in France

Increasing duties, limited means of action

François-Mathieu Poupeau

The French energy sector, very centralized and state-driven, has been transforming itself, in particular under the action of local authorities. As in many other countries (*Urban Studies*, 2014; *Energy Policy*, 2015; *Environment and Planning C*, 2017; *Renewable and Sustainable Energy Reviews*, 2020), the latter are becoming full-fledged players in the management of a field from which they have long been excluded, particularly since the nationalization of the electricity and gas industries in 1946. This greater involvement is still emerging and is certainly set to continue, in particular due to the increased use of decentralized production and digital technologies. In the meantime, the future of the governance regime remains unclear, and several interpretations coexist on the lessons to be learned from this transformation, between those who see it as a historical break and those who see it in continuity with the past.

In this chapter, we outline the new energy governance pattern that is emerging in light of current developments. To do so, we distinguish two ways in which French local authorities may intervene. The first relates to the organization of energy industries and markets. It shows that local actors still play a very marginal role in the regulation of this sector, despite the multiplication of numerous territorial initiatives. The second mode of action is about energy as a flow present in many fields that are targets of specific public policies: in the building sector (construction and renovation of housing), social action (fuel poverty), urban development, energy-climate and mobility. These areas show a much greater involvement of local authorities as a result of the decentralization laws. In a third section, we articulate these two forms of intervention (market, sectoral public policies), highlighting the many obstacles that local authorities must face, which, in our opinion, reflects a decentralization that is still timid and under strong constraint.

Local authorities and energy market regulation

The first area of local intervention is the organization and regulation of energy markets. It is historically linked to the development of the first energy networks. In France, as in other European countries, the *communes*[1] (municipalities) have been frontline players in managing these infrastructures, which initially did not extend beyond the perimeter of cities or small villages (Poupeau, 2017). This

DOI: 10.4324/9781003025962-6

resulted in the passing of a fundamental law, still in force: the law of 15 June 1906. It established a concession regime for electricity, gas and, later, heating networks. Municipalities became the organizing authorities of public energy distribution. In practical terms, they could grant a local monopoly to a company (or create their own operator) to serve their territory in exchange for a right to monitor investments (the municipality was to be covered in its entirety, according to a mutually agreed program) and tariffs (limited to maximum prices). They could do this alone or by joining forces by creating unions of municipalities (on the scale of a conurbation or a rural area).

This regime has been profoundly transformed throughout the 20th century while remaining formally in force. In this respect, a distinction must be made between two types of energy: electricity and gas, which have been taken over by large public monopolies managed by the state, and heating networks, under the responsibility of municipalities, with less state intervention.

From local to national management: electricity and gas networks

For gas and especially for electricity, the situation changed considerably from the 1920s onwards (Lévy-Leboyer and Morsel, 1994; Poupeau, 2017). Under the action of technical, economic and political dynamics, the regulation of these sectors gradually moved, in part, out of the hands of the municipalities (and their unions) and fell into those of the state. The networks were interconnected, first on a regional and then on a national scale. They became national (as opposed to local) public utilities, which gave rise to an increasing intervention by state administrations, which saw them as essential activities in the economic (competitiveness, independence) and social (access to comfort, fight against rural exodus) modernization of the country. This did not mean a radical questioning of the historic remit of municipalities in the regulation of public energy distribution (law of 15 June 1906), but rather its relativization, with generation and transmission becoming increasingly autonomous activities over which local authorities had no control. The nationalizations of 1946 and the creation of two large public monopolies, EDF (electricity) and GDF (gas), marked this new institutional order.

With the liberalization of energy markets, local authorities still play a peripheral role. As a result of the accounting and legal separation of activities previously managed in an integrated way by public monopolies (unbundling), energy supply (tariffs and services) is now outside the historical scope of the jurisdiction of local authorities, now limited to the management of "physical" distribution infrastructures. Supply is considered to be market-based, and therefore open to competition, under the watchful eye of the state (regulatory authority). Admittedly, local authorities still retain some prerogatives on the setting of regulated sales tariffs set up by the state to avoid excessive price increases for domestic consumers. But these are expected to disappear in the near future, confirming the withdrawal of local authorities. In this respect, local authorities can only act as consumers and customers for the public buildings for which they are responsible. They cannot organize purchasing groups for their inhabitants, which would result in a sort of reconstitution of the former concession regime.

However, this restriction of the municipalities' historical field of authority gets along with an opening up to new areas of action, particularly in production. This sector, formerly reserved for the two public monopolies and a few industrial players, has gradually been opened up to local authorities, which, in a liberalized market, are considered producers in the same way as other players. They (municipalities and their unions, intercommunalities, *régions* and *départements*) can thus create (semi-)public companies dedicated to the production of renewable energies in electricity and gas (see Chapter 6). But local authorities can also participate in private initiatives, including citizens. The "clean energy for all Europeans" package (2019) introduces the notion of the "renewable energy community", which allows local authorities to be involved in such projects. However, despite these new opportunities, it is clear that the role of local authorities is still marginal. For electricity, renewable energies account for only 17% of production in France, with nuclear energy still predominant[2] (see Chapter 1). Moreover, local authorities contribute only slightly to the development of these alternative energies, given the strong presence of large industrial groups. For gas, biogas production (used to generate electricity and heat, or to be injected into natural gas networks), in which local authorities are active, but again in modest proportions, remains much lower.[3] These data show the very low weight of local authorities in the French production mix.

The preservation of a stronger local influence: heating networks

Unlike electricity and gas, heating networks, mostly set up in urban areas, have always kept strong ties with municipalities. The latter can operate them either by creating local operators or by calling on private companies, often subsidiaries of large firms. The first heating networks appeared in France in the 1920s. However, their expansion phase took place mainly after the Second World War, a period of strong urbanization that led several large and medium-sized cities to build such infrastructures to serve their outskirts.[4] Often composed of large housing estates, these areas fit well to the business model of heating networks, which needs a high population density to be economically viable. As a result, the number of infrastructures increased sharply and reached 664 in 2016 for a network length of more than 5,000 km.[5]

Thanks to the responsibilities they have retained (organizing authorities or operators' owners), municipalities have, in this sector, more leeway than in electricity or gas to make their energy choices. They are more directly involved in setting tariffs, which may lead to low price policy decisions for certain population groups (however, limited by the fact that heating and cooling networks are in competition with electricity and gas). Similarly, they may try to influence the use of renewable energies in the production units that supply the infrastructure. Many conversions have been made in this way, under pressure from certain local elected representatives, ecologists in particular, to go, for example, from coal and fuel oil to biomass. At present, even if this figure cannot be attributed solely to the action of municipalities, 53% of heat production is provided by renewable energies or waste.[6] This proportion is expected to increase in the coming years, under the joint action of the state and local authorities, as part of the French strategy for the development of renewable energy.

Local authorities: emerging players but with limited powers

In the end, the industrial organization of the energy sector shows a still reduced involvement of local authorities, as summarized in Figure 3.1.

In electricity and gas, few levers currently exist, apart from public distribution, which, however, has not yet been liberalized. The introduction of competition, which is constantly being postponed, could perhaps lead to a revival of power for the municipalities, their unions and intercommunalities that are responsible for it. The situation in the heating networks appears very different, as the status of the organizing authority granted by the law of 15 June 1906 has been much better preserved. However, this observation must be put into perspective. On the one hand, the municipalities do not have total freedom of action in this sector, which is in competition with electricity and gas, as alternative sources of heat. Their degree of strategic latitude and autonomy is therefore reduced. On the other hand, heating networks have a very limited weight in energy. They currently account for only 6% of the heat and cold supply in France.[7]

Local authorities and sectoral policies

The involvement of local authorities is not confined to questions of market organization and regulation. It also concerns public policies that have a direct and indirect impact on (mainly) energy consumption, in all its dimensions, and (a little less) on supply. Historically, local authorities have first intervened as owners of their buildings, for their uses (administrative services) or for those of their population (schools, libraries, cultural and sports facilities, etc.). They have also led many actions in the management of public lighting. They continue to act in such fields as consumers. But other forms of implication have since been added, as local powers have taken on new responsibilities, in particular in the fields of housing, fuel poverty, planning and development, and mobility. In this second section, we come back to these different main responsibilities in a "chrono-thematic" way, that is, according to their order of appearance over time.

Thermal renovation of buildings

In France, local authorities have been involved in housing policies since the interwar period (Voldman, 1997). At a time when urbanization was developing at a steady pace, the aim was to promote a supply of housing that met modern hygiene standards (fight against insalubrity, comfort, etc.), while at the same time responding to growing demand. In particular, cities positioned themselves very early in

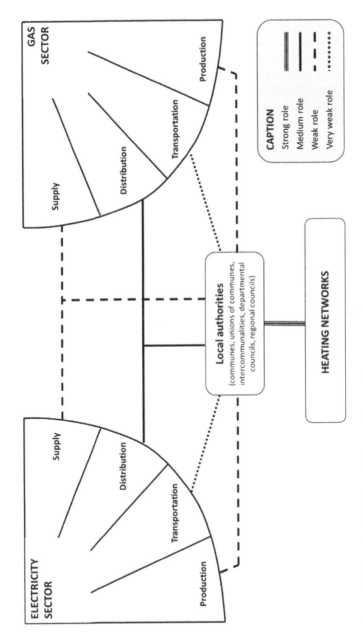

Figure 3.1 Local authorities and the regulation of the energy sector.

the production of social housing in order to enable low-income citizens to find a home under financially acceptable conditions.

In this area, interest in energy issues has been late. Until the 1970s, the major challenge for governments was to ensure massive construction of housing at low production costs (Effosse, 2003). Energy issues were certainly not neglected, but they were mainly confined to the following question: what type of energy should buildings be supplied? The question was then essentially posed in terms of energy prices. However, the production of quality, well-insulated and low-energy-consuming dwellings took a back seat.

It was only with the oil crisis of the 1970s that a greater concern arose about managing energy demand in buildings (Pautard, 2009). Under the action of the ADEME and the ANAH,[8] public policies were implemented to achieve energy savings. Thermal renovation programs began to be carried out, led by the state, which provided the majority of public funding. The commitment of local authorities on these issues, sometimes even earlier, then increased. According to a recent study, the housing sector, which represents a major challenge in terms of energy transition,[9] is today a priority area identified by local authorities.[10] Several actors, levels of intervention and instruments can be identified.[11]

Municipalities and intercommunalities are, first of all, very involved as board members of the bodies responsible for social housing management. They can thus initiate ambitious policies or provide financial support for programs to control consumption and aid renovation work. Examples include the *opérations programmées d'amélioration thermique des bâtiments* (OPATB, scheduled building thermal improvement operations) and the *opérations programmées d'amélioration de l'habitat* (OPAH, planned housing improvement operations). The action of municipalities and intercommunalities has extended more recently to the private sector, where stakes are high. Urban planning laws make it possible to impose stricter criteria for the construction of new, less energy-intensive buildings. Some cities have also created local energy agencies to raise awareness on energy issues (Poupeau, 2008). These structures, partly funded by the ADEME and the European Union, offer free advice to citizens, in particular to inform homeowners about aids to reduce the energy bill of their property.

Other local authorities are involved in thermal renovation. *Départements* can, through grants, sign partnership agreements with the ANAH and social landlords. More recently, *régions* have also taken a position on this issue. The NOTRe law voted in 2015[12] strengthens their responsibilities by making them the coordinators of energy renovation platforms. They are thus pushed to set up entities to better articulate national programs and local initiatives. Some *régions* create other tools, such as calls for projects or third-party financing companies dedicated to the renovation of buildings, sometimes in conjunction with other levels of local government.

Aid for non-payment and the fight against fuel poverty

Local authorities also play an important role in helping the most deprived populations. Far from being isolated, their action is closely linked to that of the state, which has gradually taken up the issue. As early as the 1980s, in a context of persistent economic crisis, the state compelled the public companies EDF and GDF to set up a system of aid for unpaid electricity and gas bills for the poorest people (Dubois and Mayer, 2013). Municipalities and *départements* were very quickly involved, as part of their social action responsibilities, which resulted in the integration of this aid into the *fonds solidarité logement* (FSL, solidarity housing fund). Co-financed at 50% by the state until 2005, the FSL is now entrusted to *départements*, which may delegate part of its management to the municipalities. For a long time, it has been one of the main aid tools for the most deprived populations, with the energy component accounting for the largest share.[13] Alongside it, other schemes set up by the state (aid for unpaid bills in 2004, social tariffs for electricity and gas in 2008, ban on winter power cuts in 2008, or energy vouchers in 2018) or by local authorities coexist.

The 1990s and 2000s marked a notable evolution in the care of people facing difficulties in meeting their energy needs. The first was semantic. "Management of unpaid bills", which had been used until then, gradually disappeared to be replaced by the term "fuel poverty" (Dubois and Mayer, 2013). With it prevailed a broader definition of the populations targeted by public policies. The so-called "Grenelle 2" law defined this term as follows: "is in a situation of fuel poverty a person who has particular difficulties in obtaining the energy supply necessary to meet his or her basic needs in his or her home because of inadequate resources or housing conditions".[14] Henceforth, a person is considered to be in a situation of fuel poverty when he/she spends more than 10% of his/her income on energy expenditure in his/her home. ADEME estimates the number of people concerned at 3.8 million in 2018, or 14% of French households.[15]

The other change concerned the scope of government action. For a long time, the public intervention was mainly curative and consisted in helping households to pay their energy bills. From the 1990s onwards, actions were also preventive and focused on the behavior of individuals and their housing. This paradigm shift is now reflected in more diversified strategies, both in raising awareness of energy savings and in renovating certain poorly insulated buildings. In this context, local authorities remain frontline players, particularly as part of the thermal renovation programs that they may carry out. They also intervene through the local energy agencies mentioned earlier.

From urban planning to energy-climate schemes

Local authorities also pay increasing attention to energy planning. Municipalities have been involved for a long time in planning operations, in conjunction with the state (urban modernization and development since the interwar period, policies for large housing estates in the 1950s–1970s, creation of new economic areas, etc.). But, as in the case of housing policy, energy issues were slow to be fully taken into account, usually being reduced to a trade-off between sources of supply for new urbanized areas or neighborhoods. Even the decentralization laws of the early 1980s, which transferred urban planning to the *communes*, remained silent on the subject. Pushed by EDF and GDF, the state was reluctant to strengthen local powers. This situation persisted in the years 1990–2000, despite little change. Now, municipalities and intercommunalities can carry out urban projects with a more substantial energy component, particularly in eco-neighborhoods. They may therefore influence decisions on energy supply and the extension of electricity, gas, or heating infrastructures. But such initiatives are undertaken on a fairly small scale. The power of local actors remains still limited in the rest of their territory, even if the MAPTAM law of 2014[16] has created new opportunities for metropolises (development of energy master plans).

The intervention of local authorities has been strengthening since the 1990s. It has been extending, on the one hand, to a new field of action (energy-climate) and, on the other hand, to new levels (*régions*). While previously excluded from the energy planning process, *régions* are now associated by the state to renew its practices and methods, which are considered too top-down. A first step was taken with the development of the *schémas de services collectifs* (SSC, collective services schemes) launched by the state in the late 1990s, which covered a wide range of areas, such as energy, transport, culture and health. This experience, far from being convincing, continued on other bases afterward. The fight against climate change, which has emerged as a national stake, reinforced the legitimacy of *régions*, considered as key actors to relay the strategy set by the state (reduction of greenhouse gases, deployment of renewable energies, etc.). In 2007, the *schémas régionaux climat air énergie* (SRCAE, regional climate air energy schemes) were thus created following the "Grenelle de l'environnement", a major institutionalized participatory process aimed at mobilizing all stakeholders around sustainable development. The SRCAEs were one of the new tools to achieve this (Poupeau, 2013; Gérardin, 2018; Dégremont, 2018). They were based on a closer association between the decentralized state services and *régions*, co-producing a document setting out regional guidelines for decreasing consumption, developing renewable energies, and reducing greenhouse gas emissions. In 2015, a new phase was reached with the creation of the *schémas régionaux d'aménagement et de développement durable des territoires* (SRADDET, regional schemes for the planning and sustainable development of territories), established by the NOTRe Act. By integrating the SRCAEs into a multi-sectoral approach, the SRADDETs gave a little more power to the *régions*. These schemes are more prescriptive, with regard to the other planning documents (urban planning, housing, transport). But, above all, their preparation is now entrusted to the *régions* alone.

This change in the planning process does not only affect *régions*. It also widens the scope of intervention of municipalities and intercommunalities. In particular, the latter has been increasingly promoted by the state government. In 2004, the *plans climat énergie territoriaux* (PCETs, territorial energy-climate plans) were created to rally such actors, considered as key partners for implementing national guidelines. The TECV law[17] transformed these schemes into *plans climat air énergie territoriaux* (PCAETs, territorial climate, air and energy plans). Their preparation is now compulsory for intercommunalities with more than 20,000 inhabitants. Like regional schemes, with which they must be compatible, the PCAETs are general guidance documents which, on the basis of a territorial diagnosis, identify goals and actions to be carried out with regard to the decrease of consumption, renewable energy development and reduction of greenhouse gas emissions.

The profusion of these documents, which are often produced very quickly and without any real consultation, contributes to the rise in power of local authorities at various levels. However, their effects should not be overestimated. Indeed, these schemes are very much framed by the state and the major energy operators, who do not intend to "cede ground" (Poupeau, 2013; Briday, 2020). On the one hand, the SRCAEs/SRADDETs and PCAETs must comply with national planning documents, for example, the *programmation pluriannuelle de l'énergie* (PPE, multiannual energy programming). On the other hand, they are only supposed to discuss certain issues. For example, national choices such as nuclear power generation are not intended to be addressed. Finally, these schemes do not have their own funding. Such sub-national planning responsibilities should therefore be seen as tools enabling the state to involve local actors more closely, and not as instruments for a greater decentralization (Poupeau, 2013).

Energy and mobility issues

Transport and mobility, which, along with buildings, represent a significant share of energy consumption, are among the areas of jurisdiction for French local authorities. Here too, energy issues are receiving increasing attention, since the agenda-setting of environmental (pollution) and, more recently, climate change (greenhouse gas emissions) problems. According to the survey mentioned earlier,[18] these issues raise less interest than energy renovation of buildings, public awareness, or street lighting. However, 26% of local authorities see them as a high priority and 48% as a priority.

In this area, local authorities have many levers at their disposal. First, they can act on their own assets by replacing their fleet of cars and buses with electric or natural gas-powered vehicles (NGVs), in order to eradicate the use of fossil fuels. This mode of action is becoming increasingly widespread, in particular because it demonstrates a strong political will on environmental issues. Local authorities can also rely on their responsibilities as road managers and transport organizing authorities to implement more ambitious public policies. Some are supporting soft mobility through local plans and the development of cycle paths (Huré, 2019). Others are launching schemes to deploy electromobility in their territories (Cranois, 2017; Sajous and Bailly-Hascoët, 2017). Many initiatives are aimed at installing charging stations. This foreshadows the gradual establishment of a national network able to encourage more and more drivers to use electric vehicles. In this process, cities and *régions* are increasingly active, with sometimes ambitious mobility plans. However, rural areas, through *départements* and above all energy unions of municipalities, can also set up this equipment (Cranois, 2017; Boyer, 2019).

A local rise under control

We are therefore witnessing a proliferation of initiatives and skills-taking that has been widely reported in the academic and professional literature in France. It clearly shows that the French energy model is undergoing many changes, under pressure, in particular, from decentralizing forces. In this last section, we qualify this observation by highlighting the many obstacles faced by local authorities.

Decentralization under strong political constraints

Local initiatives are first of all part of an institutional system that is still largely controlled by actors who are in favor of the historic centralized model. The first section of this chapter clearly shows that local authorities' room for maneuver remains very limited in the regulation of energy markets. The second section reminds us that the state is still very present to control local action when it comes to potentially challenging the economic and industrial order, as is the case for strategic planning.

This intervention under constraint is explained by the balance of power between supporters and opponents of decentralization. Three networks of actors are currently structuring the field of energy reforms in France on this specific issue (Poupeau, 2020a). The first ("historic Jacobins") includes key players, such as most state administrations, major energy operators, employers' lobbies and most government political parties, eager to maintain a centralized frame. The second network ("alternative decentralizers"), which is in favor of greater decentralization, brings together forces that are certainly important but emerging, including some large associations of local elected officials, pro-environmental NGOs and some political parties (especially ecologists). Between these two poles, we can identify a third network ("moderate decentralizers") made up of a powerful association of rural elected officials and a few state administrations seeking a stronger, but controlled, involvement of local authorities.

Given the positions of its members in the political, economic and administrative landscape, the network of "historical Jacobins" is currently locking the political agenda of decentralization issues. In order to limit the influence of "alternative decentralizers", it knows how to use arguments that shape the political construction of debates: promotion of the nuclear industry (as a jobs provider, an instrument of national sovereignty and a tool to decarbonize energy), support for EDF (to consolidate its status as a "national champion"), preservation of national solidarity between territories (which could be undermined by too much decentralization) and so on. Faced with it, the arguments developed by local actors still carry little weight in the national political arena, which explains their weak ability to change the rules of the game. In this context, the action of local authorities is certainly recognized. But it is rather confined to a relay function of national public policies, as for the development of renewable energies or electromobility.

Seeking financial and human levers

Local authorities are also hampered by the access to funding and a high dependence on the state. In the survey earlier mentioned,[19] 62% of respondents working for them believe that the lack of financial resources is the main obstacle to the implementation of local energy policies. They also highlight the weakness of human resources (37%) and expertise. These data echo other surveys pointing out budgetary difficulties in managing energy issues[20] and implementing the PCAET[21].

Local authorities have no specific resources dedicated to energy.[22] They must therefore dig into their own general budget to meet their ambitions. In particular, the state has, up to now, always opposed to their being able to collect part of the *contribution énergie climat* (CEC, climate and energy contribution), a tax introduced in France by a 2014 finance law to put a price on carbon. The Association of French Mayors asked that it be partly allocated to the municipalities, because of their commitment to the fight against climate change, through territorial climate schemes. In June 2018, several associations of local elected officials called for local authorities implementing PCAETs and SRADDETs to receive 10% of the 8.5 billion euros collected by the state, a sum which, according to some projections, should reach 22 billion euros in 2022.[23] For these associations, this amount would enable them to finance the drafting, organization and implementation of climate plans, estimated at, respectively, 1, 10 and 100 euros per inhabitant. Despite these pressures, the government turned a deaf ear, and the subject has been left out of the debate until today.[24] This recurring failure reflects a great reluctance of the state and, in particular, the Ministry of Economy and Finance, to entrust substantial financial resources to local authorities for energy-climate.[25]

In this context, local powers are still very dependent on state funding. New financial tools exist for them today, such as third-party financing companies or semi-public companies. However, their scope is still very limited with regard to the challenges represented, for example, by the thermal renovation of buildings and the development of renewable energies.

Local governance as a source of rivalry and fragmentation

Finally, local authorities do not present a united front that would enable them to better resist the coalition of "historic Jacobins". Many different levels of local government are involved in the energy sector, with their own vision and interest. Municipalities, historic players in the field, still retain important responsibilities. Unions of municipalities are also very present in public distribution networks, energy demand management, renewable energies and electromobility (Boyer, 2019). *Départements* keep some duties, in particular with regard to fuel poverty. Finally, *régions* have emerged as new actors in the fields of climate, air quality, energy and sustainable development. Figure 3.2, which summarizes the jurisdictions of local authorities, to which should be added "voluntary" interventions,

	Climat related jurisdictions	Crowdfunding	Production and distribution of energy		Third party financing	Energy demand management
Région	Leader (*chef de file*) "Climate, air quality, energy and sustainable development of the territory" Planning for economic development, transport, climate, air energy and biodiversity (SRADDET) Agriculture (management of European funds)		Develop distribution networks Operate a renewable energy production facility	Regional biomass and wind schemes		Public service of energy efficiency Coordination of territorial platforms for energy renovation
Département	Roads (*départementales*) Middle schools Transport of handicapped children			Energy distribution organizing authority (AODE)		Leader (*chef de file*) "Fuel poverty"
Établissement public de coopération intercommunale (municipal grouping)	Leader (*chef de file*) "Sustainable mobility and air quality" Development of several planning documents (PLU(I), PDU(I), PLH, SCoT[1], PCAET)					Coordinator of the energy transition Management of energy renovation platforms
Commune (municipality)	Leader (*chef de file*) "Sustainable mobility and air quality"					Building permit
	Roads					

Source: Réseau action climat, « Nouvelles compétences climat-énergie des collectivités territoriales », May 2016, p. 34 (according to a table from the French Ministry of the Interior).

Figure 3.2 The jurisdictions of the French territorial collectivities in energy-climate.

clearly illustrates this multiplication of levels of government. From this point of view, energy does not differ from other areas of public action in France, being characterized by an "institutional mille-feuille", which, although it may have benefits (different levels and resources can be activated by local actors wishing to tackle energy-climate issues), does not always make these interventions legible or coherent. It gives rise, in particular, to problems of competition or institutional articulation that contribute, in a way, to maintaining a centralized model.

A first example is the public distribution of electricity, one of the most important responsibilities of local authorities in the regulation of energy systems. In this field, municipalities are the organizing authorities for networks managed by operators, whether local or, more generally, national (Enedis, a subsidiary of EDF). Few have retained this function, which has been transferred either to departmental-sized unions of municipalities, mixing rural and urban communes (on a voluntary basis), or to large metropolises (a process made compulsory by the MAPTAM and NOTRe laws). Given the financial and public policy stakes that distribution infrastructure represents for these two types of local actors, this transfer gives rise to forms of competition that play in favor of the "historic Jacobins", who can rely on this rivalry to maintain their power (Poupeau, 2020b).

Energy-climate planning also illustrates the effects of this multiplication of local interventions. It shows not so much competition phenomena (even if there may be some) but rather difficult coordination between policies carried out by *régions* (through SRADDETs) and intercommunalities (through PCAETs). Although *régions* are supposed to build their SRADDETs with other local authorities (including intercommunalities), although the PCAETs must be compatible with these SRADDETs, research conducted on the subject shows a problematic articulation (Dégremont, 2018; Briday, 2020). It concerns not only the preparation of these strategic documents but also their implementation since *régions* cannot legally impose financial burdens on local authorities that would result from their own choices.

Conclusion

In this chapter, we have tried to provide an overview of local authorities' energy intervention. This is not an easy exercise, since energy is used in many areas of daily life and is involved in a large range of public policies. Consequently, the analysis of decentralization issues can give rise to very different interpretations. For some, based on the observations they can make in a few fields, the multiplication of local initiatives illustrates a break with the French historical model, marked by centralization. For others, who are interested, for example, in the organization of energy markets, the limited nature of local powers will be analyzed as a sign of continuity. It is thus difficult to draw a conclusion on the changes underway, especially in the digital age, which has not yet produced its effects and could make the general picture that we can draw a little more complex.

Despite these uncertainties, it seems to us that the decentralization of the French energy sector is still in its early stages. We have discussed the main reasons for this. The first one is political and institutional since the dominant energy coalition is still very much in favor of centralization and does not seem to be really challenged, at least in the short term. The second reason is financial, insofar as local authorities still have little room for maneuver, being caught up in a system of resource allocation – not specific to energy – which strongly constraints them. The third main reason is specific to the local governance of energy issues and the multiplication of actors that can hardly articulate their intervention in order to offer a "united front" against the advocates of centralization. These are as many reasons – to which one could add others – that invite a cautious reading of the current transformations, the term "local involvement" being, at this stage, more relevant than "decentralization" to describe the situation.

Notes

1 About the territorial organization of France, see the introduction of this book.
2 Source: Ministère de la transition écologique et solidaire (MTES).
3 In 2017, biogas accounted for only 2.3% of gross electricity production from renewable sources and 2.8% of primary heat consumption. Moreover, in 2017, very little biogas (3%) was injected into gas networks via its transformation into biomethane (CGDD (2019) *Chiffres clés des énergies renouvelables 2019*, pp. 8 and 53). Available at: www.statistiques.developpement-durable.gouv.fr/sites/default/files/2019-05/data-lab-53-chiffres-cles-des-energies-renouvelables-edition-2019-mai2019.pdf (Accessed: 24 March 2020).
4 CEREMA (2018) *District heating and cooling in France*. Available at: http://reseaux-chaleur.cerema.fr/district-heating-and-cooling-in-france (Accessed: 24 March 2020).
5 Ibid.
6 Ibid.
7 CEREMA, op. cit.
8 ADEME: Agence de l'environnement et de la maîtrise de l'énergie; ANAH: Agence nationale de l'amélioration de l'habitat.
9 It should be remembered that, in France, the building sector is the largest source of greenhouse gas emissions behind transport (source: MTES).
10 Study carried out from 4 December 2019 to 8 January 2020 by Infopro digital études for the newspaper *La Gazette des communes* on a sample of 350 local authorities (20% elected officials and 80% agents) ("Transition énergétique: des collectivités bien engagées mais des moyens insuffisants", *La Gazette des communes*, 25 February 2020); 46% of respondents indicate that energy renovation of buildings is a high priority, 40% that it is a priority. Available at: www.lagazettedescommunes.com/663723/transition-energetique-des-collectivites-bien-engagees-mais-des-moyens-insuffisants/ (Accessed: 24 March 2020).
11 For a fairly recent review, see *Plan bâtiment durable* (2017) "Financements de la rénovation énergétique des logements privés et déploiement du tiers-financement : état des lieux et perspectives", Rapport remis à la Ministre du Logement et de l'Habitat Durable. Available at: www.planbatimentdurable.fr/IMG/pdf/170321_rapport_financements_de_la_renovation_energetique_des_logements_prives_et_deploiement_tiers-financement-2.pdf (Accessed: 18 March 2020).
12 Loi n°2015-991 portant nouvelle organisation territoriale de la République (law on the new territorial organization of the Republic), passed on 7 August 2015.
13 About 83 million euros in 2010, which then concerned approximately 328,000 households (Assemblée des départements de France, Ministère du logement, de l'égalité des territoires et de la ruralité, "Place et rôle des Fonds de Solidarité pour le Logement (FSL) dans la politique sociale du logement: état des lieux et perspectives", rapport final 2015, p. 75). Available at: www.gouvernement.fr/sites/default/files/contenu/piece-jointe/2016/02/etude_fsl_et_courrier.pdf (Accessed: 18 March 2020).
14 Article 1-1. de la loi n°2010-788 portant engagement national pour l'environnement (law on a national commitment to the environment), 12 July 2010 (translated from French).
15 Available at: www.ademe.fr/expertises/batiment/quoi-parle-t/precarite-energetique (Accessed: 17 March 2020).
16 Loi n°2014-58 de modernisation de l'action publique territoriale et d'affirmation des métropoles (law on the modernization of territorial public action and affirmation of metropolises), passed on 27 January 2014.
17 Loi n°2015-992 relative à la transition énergétique et à la croissance verte (law on energy transition and green growth), passed on 17 August 2015.

18 "Transition énergétique: des collectivités bien engagées mais des moyens insuffisants", *La Gazette des communes*, op. cit.
19 Ibid.
20 In 2017, only 19% of municipalities had a dedicated agent, compared with 51% for intercommunalities (enquête ADEME communes, 2019, p. 37).
21 The *Assemblée des communautés de France* (ADCF), which brings together intercommunalities, identifies, in 38% of cases, the "human resources" factor as a brake on the adoption of PC(A)ETs (Bosboeuf, Dégremont and Poupeau, 2015).
22 Apart from local electricity taxes, that can only be collected by (unions of) municipalities and *départements*, or royalties perceived on concession contracts.
23 "Les collectivités vont-elles enfin bénéficier des recettes de la taxe carbone?" *Actu Environnement*, 19 June 2018. Available at: www.actu-environnement.com/ae/newsletter/newsletter_quotidienne.php?id=2095 (Accessed: 25 March 2020). "Les collectivités invitent le gouvernement à travailler sur la redistribution de la TICPE", *Environnement Magazine*, 19 June 2018. Available at: www.environnement-magazine.fr/territoires/article/2018/06/19/119729/les-collectivites-invitent-gouvernement-travailler-sur-redistribution-ticpe (Accessed: 25 March 2020).
24 "Contribution énergie climat: les territoires vont-ils obtenir gain de cause?", *La Gazette des communes*, 13 July 2018. Available at: www.lagazettedescommunes.com/573561/contribution-energie-climat-les-territoires-vont-ils-obtenir-gain-de-cause/ (Accessed: 25 March 2020). Questioned a few months later by Senator Guillaume Gontard, Agnès Pannier-Runacher, Secretary of State to the Minister of the Economy and Finance, simply eluded the subject. Available at: www.senat.fr/questions/base/2018/qSEQ18110535S.html (Accessed: 25 March 2020).
25 Testimony collected during a meeting organized by the association Amorce (which namely defends the interests of heating networks) about SRADDETs and PCAETs (9 March 2018).

References

Bosboeuf, P., Dégremont-Dorville, M. and Poupeau, F.-M. (2015) 'Les Communautés et les politiques énergie-climat en France. Quelques enseignements autour d'une enquête de l'ADCF' in Marcou, G., Eiller, A.-C., Poupeau, F.-M. and Staropoli, C. (eds) *Gouvernance et innovations dans le système énergétique. De nouveaux défis pour les collectivités territoriales?* Paris: L'Harmattan, pp. 121–149.

Boyer, M. (2019) 'Les syndicats d'énergie: bras armés des collectivités territoriales dans la transition énergétique?', *Working Paper LATTS*, 19–17. Available at: hal-02166300.

Briday, R. (2020) 'Le SRCAE d'Île-de-France. Ou l'épineuse territorialisation des objectifs nationaux de transition énergétique', *Working paper LATTS*, 20–18. Available at: hal-02922745.

Cranois, A. (2017) *De l'automobilité à l'électromobilité: des conservatismes en mouvement? La fabrique d'une action publique rurale entre innovations et résistances.* PhD Thesis, Université Paris-Est.

Dégremont, M. (2018) *Transitions énergétiques et politiques à l'orée du XXᵉ siècle. L'émergence en France d'un modèle territorial de transition énergétique.* PhD Thesis, Sciences Po Paris.

Dubois, U. and Mayer, I. (2013) 'La problématique de la précarité énergétique: un état des lieux franco-allemand', *Droit et Gestion des Collectivités Territoriales*, 33, pp. 247–256.

Effosse, S. (2003) *L'invention du logement aidé en France: l'immobilier au temps des Trente Glorieuses.* Paris: Comité pour l'histoire économique et financière de la France.

Energy Policy (2015) 'Urban energy governance: Local actions, capacities and politics', Special issue, 78.

Environment and Planning C (2017) 'Sub-national government and pathways to sustainable energy', Special issue, 35(7).

Gérardin, N. (2018). *Vers une centralité de la Région? Émergence et affirmation du rôle de la Région Île-de-France en matière climat-air-énergie*. PhD Thesis, Université de Paris-Saclay.

Huré, M. (2019) *Les mobilités partagées. Régulation politique et capitalisme urbain*. Paris: Éditions de la Sorbonne.

Lévy-Leboyer, M. and Morsel, H. (eds) (1994) *Histoire de l'électricité en France. Tome deuxième: 1919–1946*. Paris: Fayard.

Pautard, É. (2009) *Vers la sobriété électrique: politiques de maîtrise des consommations et pratiques domestiques*. PhD Thesis, Université de Toulouse 2.

Poupeau, F.-M. (2008) *Gouverner sans contraindre. L'agence locale de l'énergie, outil d'une politique énergétique territoriale*. Paris: L'Harmattan.

Poupeau, F.-M. (2013) 'Quand l'État territorialise la politique énergétique. L'expérience des schémas régionaux du climat, de l'air et de l'énergie', *Politiques et Management Public*, 30(4), pp. 443–472.

Poupeau, F.-M. (2017) *L'électricité et les pouvoirs locaux en France (1880–1980). Une autre histoire du service public*. Bruxelles: Peter Lang.

Poupeau, F.-M. (2020a) 'Everything must change in order to stay as it is. The impossible decentralization of the electricity sector in France', *Renewable and Sustainable Energy Reviews*, 120. https://doi.org/10.1016/j.rser.2019.109597.

Poupeau, F.-M. (2020b) 'Metropolitan and rural areas fighting for the control of electricity networks in France. A local geopolitics approach to energy transition', *Environment and Planning C: Politics and Space*, 38(3), pp. 464–483.

Renewable and Sustainable Energy Reviews (2020) 'Energy decentralization', Special issue, 120.

Sajous, P. and Bailly-Hascoët, V. (2017) 'Électromobilité: acteurs, structurations en cours. Étude à partir du cas haut-normand', *Recherche Transports Sécurité, NecPlus*, 01–02, pp. 1–21.

Urban Studies (2014) 'Urban energy transitions: Places, processes and politics of socio-technical change', Special issue, 51(7).

Voldman, D. (1997) *La Reconstruction des villes françaises de 1940 à 1954. Histoire d'une politique*. Paris: L'Harmattan.

4 Local energy governance

The Japanese context, development and typology

Aki Suwa

There is an increasing recognition that energy transition has become the global climate imperative. Although the energy transition is primarily policy-driven at international and national levels, participation of citizens in local energy decisions is also key to delivering successful energy transition: in fact, energy transition holds a particular significance in local and regional boundaries, as it addresses energy and economic interaction, as well as collective interests among the participating individuals in a set of spatial borders.

The arguments over local energy decisions have been changed from simply arguing how to deliver regional reconciliation between supporting and opposing opinions to renewable development, to how to distribute local benefits (Walker et al., 2007). Japan also follows international footsteps where, in the early debates, the primary focus was on negotiations and compromises between different stakeholders in a spatial boundary on the development of locally available renewable resources. The discourse of community energy was then advocated as a response to the need to reconcile divisive interests, where community energy development has become the platform or project that incorporates community-led, -controlled and -owned renewable energy installation development.

In undertaking the observation of international practices, the energy-related benefits are increasingly connected to economic systems in which equity in production, distribution, benefit sharing and equal investment opportunity hold the key to a regional sense of integrity (Clausen and Rudolph, 2020; Rudolph et al., 2014; Aitken, 2010; Allan et al., 2011). Based on these arguments, it is crucial to understand what and how local hegemony is developed in order to help increase community interests and trust. Benefit sharing generally has an aim to establish mutually positive outcomes, where different perceptions for equity allocation could have a strong influence over the project outcome. Special attention is paid to the fact that the benefits are not necessarily directly financial, but could be allocated through local employment, revenues from contracts, margins from co-ownership and tax incomes to local government (Walker et al., 2014; Bellaby, 2010).

The shifting arguments have thus gradually materialized as the development of new types of local energy governance. Recently, numerous examples are emerging internationally to establish and re-structure local and regional institutions to develop renewables in a collective manner (e.g., energy cooperatives) (Schmid,

DOI: 10.4324/9781003025962-7

2019). These institutional developments involve, however, not just a mere organizational establishment, but also the establishment of a governance mechanism, which integrates the increasingly complex myriad of stakeholders through the re-arrangement of ownership architecture, where the spatially overarching communal benefits and resources are at stake.

As energy governance generally refers to "the rules, processes, practices and behavior that affect the way in which energy is generated and used" (Wade et al., 2013), there is now a need to shift the governance perspective from a centralized structure to a more decentralized and local scale, where increased participation and commitment from local government and communities into energy decisions are expected. In this context, energy governance holds a profound significance in delivering resource and economic autonomy, while ensuing benefit sharing among community members. The discussion of benefit sharing suggests that local energy governance shall have a more progressive role than the mere local version of the state energy governance, ensuring social innovation, invoking changes in actor configurations and resource access within the energy system that seeks to achieve low carbon goals while increasing the general wellbeing of communities (Warbroek, 2019; Hoppe et al., 2015).

This chapter aims to explore the multiple concepts and organizational arrangements materializing the institutional change and to discuss what these can mean for community welfare and local energy governance. Using case studies in Japan, this chapter aims to address how discourses influence the emergence of local energy governance and what the enabling factors behind this emergence are. It also highlights the implication of local energy governance in relation to energy security on a national scale. It pays particular attention to categorizing types of governance and organizational structure, and in mapping out a variety of organizational and ownership governance archetypes, which would become the basis for assessing the degree and scope of local governance in delivering community benefits. The concluding section summarizes the main findings and interpretations by reflecting on the implications of the current observations.

Connecting local energy governance with national wealth and benefits sharing discourses

National wealth and resource security

One of the major challenges for climate change is to shift from the dependence on fossil fuels and irrevocably establish a society with more sustainable and renewable energy use. The Paris Agreement within the United Nations Convention for Climate Change (UNFCCC) acknowledges that climate change is "a common concern of humankind", urging parties to strengthen the global response to the threat of climate change. It explicitly requires holding the increase in the global average temperature to well below 2°C above pre-industrial levels and pursuing efforts to limit the temperature increase to 1.5°C above pre-industrial levels

(UNFCCC, 2016). For achieving this, the significance of securing sufficient indigenous non-fossil energy resources has become even greater than ever.

In addition to the climate emergency, international conflicts and the associated geopolitics, observed in history and the contemporary global landscape, need to be addressed in the context of renewable energy development. A country's military and economic dominance in the international order has been repeatedly determined by access to strategic energy resources, where oil counts as the priority. As in the historical event of the oil crisis in the 1970s, vulnerability related to fossil fuel dependency has been witnessed in many countries. Overall, the lives and economies of modern developed countries are continuously affected by the availability of energy resources and associated technologies. With this background, especially in countries with fewer fossil reserves than Japan, it has long been argued whether and to what extent foreign energy resource dependency should continue.

Those seeking alternative energy resources and technology, however, face a mixture of policy directions. On one side, there is regulatory guidance to encourage renewable energy development. The increasing adoption of feed-in tariffs, a mechanism to support cost-based prices for renewable electricity for members of the international community is a typical example of strategic impetus to support renewables. At the same time, however, the fluctuation of international oil price levels, and fossil fuel subsidies made available by many countries, often mask and undermine the significance and urgency for substituting fossil fuels with renewables. In many countries, oil prices have been strategically lowered by tax breaks and subsidies to the extent that they undermine international efforts to prevent climate change, representing a drain on national budgets. The Organisation for Economic Co-operation and Development (OECD) estimated its members have spent US$55–90 billion a year supporting fossil fuels (OECD, 2012).

Japan, inter alia, is one of the heaviest importers of oil, as its fossil fuel-related national outflow accounts for about 30% of the total national export charges (Taghizadeh-Hesary et al., 2017). The national outflow is surely the accumulation of regional expenditure on fossil fuels, suggesting that the financial resource would emigrate when the industry, transport and household sectors in their respective regions consume energy of fossil origins, while opportunities for green investment and finance may stay slim.

Compared with imported fossil fuels, renewable energy is usually produced within national boundaries (especially "island" countries as Japan). This domestic characteristic of renewables should be extensively recognized in the context of global macroeconomics. While greater awareness is growing that renewable energy plays a significant role in national energy security, how local energy supply can contribute to increasing national energy security has yet to be fully explored. Arguments can be made as to whether and to what extent domestic renewables can save on the cost of fossil fuels, enhancing the wealth and welfare prospects of citizens by reducing fossil-related national expenditure. At the same time, since the national account is ultimately the gross sum of regional and local economic activity, community decisions on renewable use can determine the degree of statewide fossil fuel-related monetary outflow.

At the theoretical and pragmatic level, the notion of national wealth has been heavily researched since the 17th century. National wealth is the aggregate of the values created by nature and human resources through the means of production, as the property of citizens belonging to the country (Hamilton and Hepburn, 2017). The natural resource dimension, however, has been less considered in the discourse of national wealth, where only recently national wealth was defined as the potential of socio-economic development of a country, which is a set of resources available in the state where the resource factor is taken into account (ibid.).

In addition to the physical natural resource aspect, resource implication is gradually connected to the macroeconomic management argument. For example, macroeconomic mechanisms (e.g., sovereign wealth funds) are envisaged as the mechanisms to adjust national wealth balances, long affected by the volatility of resource revenues and expenditure (Atkinson and Hamilton, 2020). In such an argument, a missed opportunity is highlighted at the macroeconomic level, but microeconomic wealth at local and regional scales is also examined, given the aggregate nature of the micro influence to the macro scale. As much as national wealth functions as a basis for addressing an opportunity for distribution, regional and local wealth act the same way to deliver socio-economic needs on a communal scale. Thus, the reflective expectation and implication should be attached to local energy governance, which foresees and manages the communal distribution of wealth created through natural endowment, which, in turn, cumulatively affects the aggregate of sovereign wealth.

Benefit sharing and local energy governance

This interaction between national and local wealth demands a debate on how to translate the increased national income, by reducing importing fossil resources, into alternative forms of productive capital. There is argument on the so-called resource curse, where countries living primarily off their exports of natural resources do not follow a sustainable path of capturing and accumulating wealth. The problem is equally relevant to the countries that mainly import natural resources, where importing business entities accumulate profit out of the resource trade, but often consumers drain their financial resources in exchange. Therefore it is important in asking the question of appropriate measures of net savings for countries such as Japan that are large natural resource importers (ibid.), and more critically, how the savings are to be allocated into localities to optimize the associated benefits to the people on the ground.

Local energy development is expected to play an important role both in economic and social integration. Japanese rural areas, for example, are facing rapid aging and community deterioration problems. That means local revenue through household and business taxes decline, to the extent that the quantity and quality of local government services may become difficult to maintain. Whereas facing financial difficulty, as Figure 4.1 demonstrates, energy-related expenditure

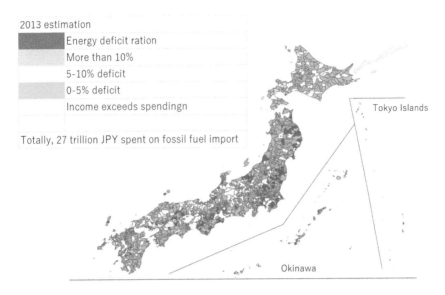

2013 estimation

Energy deficit ration
More than 10%
5-10% deficit
0-5% deficit
Income exceeds spendingn

Totally, 27 trillion JPY spent on fossil fuel import

Tokyo Islands

Okinawa

Figure 4.1 Local income deficit. (Source: Ministry of Environment Japan, 2017. Adapted
by author)

exceeds gross municipal development, suggesting many municipalities spend
more on energy than they earn from their economic activities.

There is a great expectation attached to locally produced electricity and the associated income to enhance community financial viability. A number of local energy
governance initiatives are evolving in response to this requirement. The regional
and local governance, however, are often difficult to emerge on their own without
sufficient and effective enabling stimuli. Regions may often lack convincing visions
on available governance options, as to what framework could bring benefits to a particular locality (Walter, 2014, Cowell, 2007; Cowell, 2011). In other words, there
is a need to map out the types of local energy governance models to help localities
develop appropriate renewable schemes and provide a clear understanding of what
issues are involved and contextualized behind each governance arrangement.

The issue of increased integrity and well-being of communities is a critical factor
behind the emergence and design of local energy governance, where both internal
and external factors influence the effective design of governance. There has been
a volume of literature trying to identify external and internal factors (Warbroak,
2019; Wade et al., 2013), where the external factors broadly suggest the degree to
which the community has linkages and access to the government and intermediar-
ies assuring policy and support, whereas internal factors are addressed as factors
relating to human and capital and physical resources, interaction with the local
community and linkages to government and intermediary institutions (Warbroak
et al., 2019). Largely, a new approach to incubate a good governance arrangement
is required upon emerging decentralized electricity systems (Brisbois, 2020).

Institutional change is of particular significance as a prerequisite for local energy governance. If the energy market is dominated by state-centered decisions, it is difficult to see the emergence of local energy resolutions at all. If, on the other hand, the energy market is progressively liberalized, as seen in many countries, the processes of re-configuring energy infrastructures and services affecting the municipal governance overseeing "energy commons" could become possible.

The energy markets in Japan, formerly highly centralized, are now open to new entrants. The question is whether local energy governance has emerged and the key factors in creating community benefits. This chapter focuses on Japan. First, it sets out the institutional energy challenges for national and local layers, then it analyzes some cases through the lens of benefit sharing arrangement and assesses the scope of local energy governance.

The strategic challenge of the Japanese energy transition

Historical background and policy development

Historically, a number of municipal and communal utilities existed in Japan before World War II. Prior to 1939, the Japanese electricity system was principally the auspices of local authorities and small energy entities, where these organizations were engaged in electricity production and the supply to local consumers. The size and types of municipal and communal electricity utilities varied, but most of them were using hydropower, though relatively small in scale, to provide electricity to specific localities.

These municipal utilities were, however, merged into a centralized electricity utility by the Japanese pre-war government in order to consolidate strategic resources. This was a national war-time mechanism to supply electricity to prescribed ends.

The centralized structure remained for a while after the Japanese defeat in World War II until the US occupation government arranged a horizontal separation of the centralized electricity entity, resulting in the establishment of nine (later ten) regional monopolies in 1951. This arrangement allowed each utility to own and operate plants and distribute electricity to their captive consumers (Kikkawa, 2012).

The 1950s also marked a historical turning point in Japanese fuel policy. Urged by the Cold War, the United States radically shifted its policy to allow Japan to have wider access to petroleum resources. Hydropower dominance in the Japanese electricity production structure, seen before World War II, was turned into an oil-intensive one. Notorious Japanese pollution incidents (e.g., the Yokkaichi Asthma case) in the 1950s and the 1960s are, at least partly, the result of a rapid intensification of petrochemical industrial activities. As can be seen in Figure 4.2, the dominant electricity source was changed from hydro to fossil fuels within a very short period, while the fossil intensity remained until as recently as in 2020.

Moreover, since the 1980s, electricity market liberalization processes have been witnessed in the international context. In Japan, the first wave of liberalization

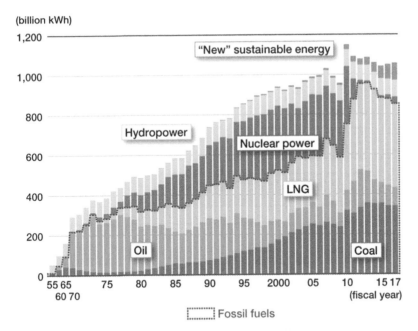

(billion kWh)

Figure 4.2 Japanese electricity source transition. (Source: Nippon.com, 2019)

started in 1995. The power market gradually became open to competition for middle- to large-sized customers, with the initial goal to create a competitive environment to reduce energy prices (METI, 2013).

The process, however, was suspended during the 2000–2010s. Then, the Fukushima nuclear plant accidents in 2011 re-activated the argument, with a reflection that utilities in nuclear investment may have been disproportionally favored in a monopoly landscape. Since 2016, the market has been fully opened to allow about 85 million households and small businesses to choose their electricity suppliers (Suwa, 2020) (Figure 4.3).

The retail market liberalization was followed by the unbundling of the ten utilities into electricity production, transmission and retail entities in order to secure the level playing field between the incumbent and newly established utilities. However, Japanese unbundling is primarily "functional unbundling", which means the transfer of operational functions and transmission and distribution divisions to separate organizations, not a legal unbundling that seeks to separate the transmission and distribution divisions into separate companies.

The liberalization in Japan has had mixed effects: on one side, it has certainly opened up a possibility for expanded parties to venture into power production. There are over 600 business entities that have started to supply electricity directly

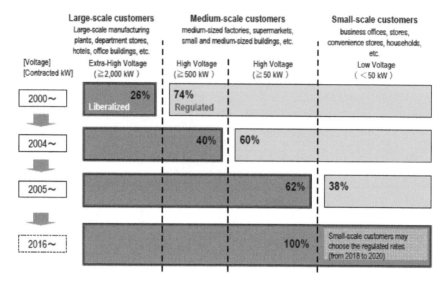

Figure 4.3 Japanese electricity market reform. (Source: TEPCO)

to customers. Many of them appeal to customers with electricity price discounts, for example, packaged concessions with telephone and mobile packet subscriptions. In contrast, some new entrants are mindful of cooperative energy development models, actively promoting decentralized renewable electricity.

Among these new entrants, there are a number of local governments, groups and businesses that have started to develop renewable electricity using resources available to their community. Also, several local governments established power producer suppliers (PPS), often with the cooperation of local stakeholders, to develop and supply renewable energy sources available to their community.

On the other hand, at least on the consumer side, the process has had a limited impact on market liquidity, with fewer than 5% of customers having so far changed suppliers (OCCTO, 2017). This may be a result of Japanese customers' conservatism over the choice of suppliers, lack of information as to energy and sustainability issues and/or the marginal financial and other benefits to customers upon changing suppliers.

Renewable policies: renewable portfolio standards to feed-in tariffs and community energy development

Japan is largely dependent on fossil fuels, which account for over 70% of total power generation. The Great East Japan Earthquake on 11 March 2011 became a background to the high dependency on fossil fuel since over 50 nuclear power plants were ordered to suspend their operations. Nuclear share within the generation portfolio of utilities decreased while LNG and coal power generation were used to make up the electricity deficiency, forcing utilities to import fossil fuels (Figure 4.1).

Renewable energy deployment has been on the Japanese government energy policy agenda for, at least, decades. After the oil crisis in the 1970s, a significant amount of the government budget was allocated to renewables research and development. The Japanese government initiated a series of projects to support renewable technologies. Its primary focus, however, was mainly on technology research and development, while less attention was paid to public policy to support and deploy renewables (Suwa and Jupesta, 2012).

Japan first introduced Renewable Portfolio Standards (RPS) in 2003, but they turned out to be ineffective in increasing renewable penetration (ibid.). The unproductive RPS was replaced with the feed-in tariff (FIT) system, adopted by law in 2011 (ibid.). The Japanese FIT accelerated renewable development, PV particularly, and increased capacity (Figure 4.4) to the extent that grid parity has been largely achieved, where the cost of PV electricity production became less than its price when purchased from the utilities (Figure 4.5) (Kimura, 2015). With this background, the government is considering substituting PV FIT with feed-in premium (FiP), the system where renewable electricity producers receive a premium on top of the market price of their electricity production.

Typically, FiP can either be fixed at a constant level regardless of market prices or change depending on the fluctuation of market prices, but the Japanese

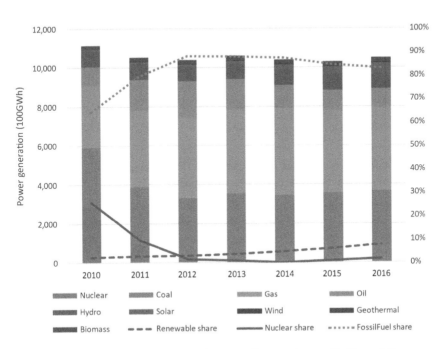

Figure 4.4 Trends of power generation in Japan. (Source: Agency for Natural Resources and Energy. Figure created by author)

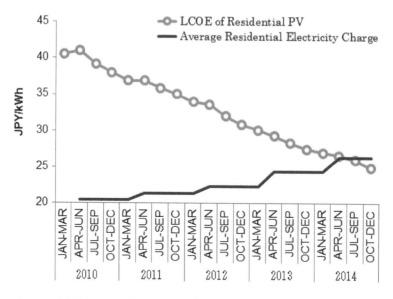

Figure 4.5 Grid parity achievements. (Source: Kimura, 2015)

government has been seen to choose the sliding FiP. Only PV is the target of policy change, whereas the other resources remain under the FIT framework (Figure 4.6).

The policy design of PV-FiP has not been decided at the moment of writing (July 2021), thus there are uncertainties on the future direction of renewable policy, where the stakeholders are cautiously observing how the policy change may influence the profitability of PV development. Nevertheless, FIT for other resources than PV remains the main renewable policy mechanism that decides the feasibility of renewable projects.

Local energy governance development: the emergence of new organizational archetypes in Japan

The interest in energy income from renewable sources has gradually paved the way for a number of "communities" to venture into the development of locally accessible energy resources. Electricity liberalization and FIT now play a significant role, as it creates new stakeholders, for example, local communities, which can enter into the market as energy "producers". Currently, the electricity market itself has a JPY 60 trillion sales volume in Japan as a whole. If local communities successfully establish energy companies with sufficient customers, they can take over part of the profits that may have otherwise been accrued to existing utilities.

These contexts found an expression in the emergence of new organizational archetypes at local and regional layers in Japan. The prominent instances include

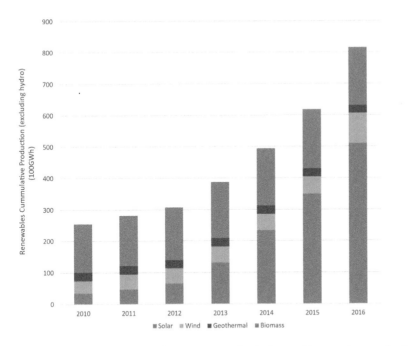

Figure 4.6 Trends of renewable energy capacity in Japan. (Source: Agency for Natural Resources and Energy. Figure created by author.)

a cooperative model, establishment of community business entities and forming municipal energy utilities. Although this emergence of archetypes is greatly inspired by the preceding organizational arrangements seen in many European countries, there are differences, as well as commonalities, with the European examples. On the basis of three examples, we observe different types of local government entities in Japan and highlight how these institutional arrangements took place and the impact they had on energy transition.

Cooperative model

A cooperative model generally refers to a scheme where citizens collectively own and manage renewable energy projects at the local level (Bauwens et al., 2017). Though there are variations in detail, they are essentially the mechanisms to produce renewable energy using local resources, and then distributing the profits earned from renewable production among the members of the cooperative depending on their contribution to the project.

In Japan, there are more than 1000 cooperatives with around 60 million members. The origins of Japanese cooperatives date back to the early 20th century. Japanese cooperatives obtained a distinct feature of the collective purchase of commodities (e.g., foods and other daily necessities) to address the co-development of agricultural and local autonomy from around the 1970s when pollution incidents threatened the safety of diets. Most Japanese cooperatives today have their own network of agricultural and other "environmentally conscious" product supply chains, with a system of transportation to deliver the goods to their members' doorsteps. It means there are extensive connections between the producer and consumers that are independent of the conventional supply chain owned and operated by incumbent retail-market players.

The first "energy" cooperative model was envisaged in Japan (Shiga prefecture) in 1996 when a group of local residents initiated a project to develop communal rooftop photovoltaic facilities. The initiative was largely emulated from a similar case in Germany, where residents in Aachen, northwest Germany, addressed the prototype feed-in tariff in 1995. The Shiga model was a combination of a residential initiative to install PV facilities and the support of the Shiga prefectural government, who purchased the produced electricity at a fixed price for 20 years. The Shiga model was followed by a number of similar initiatives across Japan.

The weight of the renewable cooperative sector, however, has shifted gradually since the early examples of shared electricity production to a more institutional arrangement of benefit sharing. It was felt strongly that there should be a sturdier institutional arrangement to let the financial profits stay within the community by redirecting the monetary outflow associated with residential electricity usage. Here, the paradigm shifted from profit distribution of energy production to a progressive mechanism to collect retail electricity income and spend it on communal purposes such as education, elderly services and local public transport (Figure 4.7).

The emerging recognition let cooperatives gain the status of electricity retailers, as the market regulation allowed them to do so, and the cooperatives often use the electricity sales profit to develop and purchase renewable energy within their

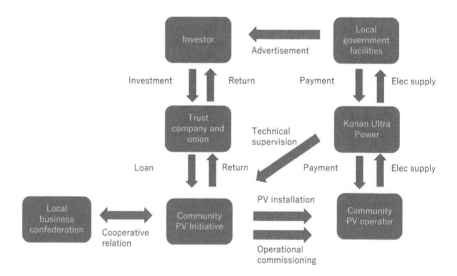

Figure 4.7 Konan Ultra Power scheme.

communal areas. In Shiga, a quasi-cooperative electricity utility (Konan Ultra Power) was formed, and several cooperatives similarly initiated electricity sales to their members (Hokkaido, Nara, etc.). In this scheme, local cooperatives form another layer of institutional arrangement, which is imperative to accumulate and share local benefits.

Community business entities

The cooperative model, both as a producer of local renewable energy and a supplier of the produced energy, addresses a progressive arrangement in using the governance resources already available in Japanese localities. They, however, have limitations in the scope of stakeholders, as the cooperatives are owned and operated primarily by their members, and the beneficiaries are naturally restricted within the membership. A new arrangement is required when a new mechanism of benefit sharing is sought, regardless of the availability of cooperative resources.

This section focuses on a case in Kyushu, the southeast part of Japan, where the financial and other benefits of renewable (geothermal) electricity development are accumulated and shared through the establishment of a local business entity. It then sees how and what benefits to the local community were created, and how that can convince localities to accept and undertake new developments.

- *Case: Environmental and social consideration for geothermal developments: Waita village, Kumamoto*

The Waita area is a small village in Kumamoto prefecture, on the Japanese island of Kyushu (Figure 4.8). It established a local group in July 2012, which was registered as a joint company in 2016. The joint company is a corporation that legally requires the company members to be limited to local residents. The new company, in collaboration with an energy engineering company, developed geothermal flash electricity of 2 MW and started selling electricity to gain financial return through the national feed-in tariff system in June 2015.

Back in 1996, there was a 20 MW geothermal development project planned by a big company coming "outside" of the locality. Waita village had a large debate as to whether they should accept the project. In the end, the plan was rejected on the ground that there were not sufficient legal and financial measures against any unforeseen geological repercussions. The villagers were split between for and against the project with bitter feelings left.

Against this background, the joint company project emerged as a way to reunite the village and avoid divisions. The participating 26 villagers contributed JPY 10,000 (US$100) each to create a fund to establish the Waita Community Joint Company. Though rather modest in the capital size (originally JPY 260,000, US$2,600), it succeeded in finding an energy engineering partner (Chuo Electric Power Furusato Heat and Power, CEPFEP), to build a geothermal flash power

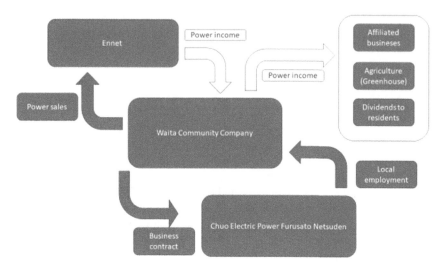

Figure 4.8 A business scheme in Waita, Kumamoto.

plant, and it commissioned the partner to "build, own and operate" the power plant.

For the new scheme, it was agreed that the electricity income would be divided by the Waita Joint Company and CEPFEP with a ratio of 20%:80%. Though the share is higher for the CEPFEP, it trained and hired local villagers to be company members, thus their salaries circulated back to the community. Also, CEPFEP has ensured its technical headquarters was located and registered at the village, meaning the tax income is generated to the local authority.

The villagers, who were initially skeptical of geothermal development, became technically trained and had an increasingly strong sense of ownership of the geothermal facilities. The financial benefits returned to the villagers in the form of direct dividends and salaries, and the accrued funds are going to be used to develop a second geothermal energy plant, in addition to local agricultural facilities. Near the power plant, there are about 100 hectares of forestry commons, which the villagers regularly maintain. This eases the installation of monitoring posts to inspect geographical and environmental fluctuation in the vicinity.

The Waita Community Joint Company is now planning its second geothermal power plant, although it may face difficulty settling grid connectivity issues. Also, there is an unforeseen future after 2030 when the current FIT deal of JPY40/kWh electricity sales will expire. The company is keen therefore on the development of additional income through cultivating crops and using geothermal residue heat, with higher market value. The Waita case demonstrates an important implication in securing financial benefits for the community (Suwa and Sando, 2018).

Forming of municipal energy utilities

There are local authorities that have started to develop renewable electricity using resources available to their community. These community-based renewable electricity developments are envisaged to play an important role both in economic and social integration senses. As touched upon, Japan is facing rapid aging and community decline problems, which means that decreased income tax revenue is plausible, and there are increasing expectations that locally produced electricity will become a tool to gain financial income through the sale of generated electricity. It would then create capital that can benefit the localities. At the moment of writing, there are about 60 PPSs in Japan, either established or invested by local authorities. The following section will focus on a case in the city of Izumisano, Osaka prefecture, Japan, to illustrate how municipal energy business can generate local benefits.

- *Case: Izumisano Electric Company (Izumisano PPS)*

Izumisano city is located in the south part of Osaka prefecture, Japan, with a population of about 100,000. The city was founded in 1948 and was for long a suburban city to the historical central Osaka.

Its fate changed when, with another two neighboring local governments, Kansai International Airport was established there, allowing the construction of a manmade island. The city government expected a wave of inbound tourists to flow into the city with the opening of the Kansai International Airport in 1994. It invested heavily in various airport-related projects, including the Rinku Town development. The Japanese economy bubble burst in the 1990s, reducing financial expectations, and any return of the investment became virtually hopeless. In 2008, the deficit level of the Izumisano City Government (ISC) was 24%. It became a listed "financially unstable local authority" by the national government.

After initiating various contingency actions, the balance sheet of ISC improved around 2010. As part of its plans to recuperate the financial situation, ISC envisaged the establishment of the Izumisano Power Producing and Supplying Association (Izumisano PPS), together with Power Sharing Co., Ltd (a technical co-business entity with its headquarters in Tokyo) in 2015 with the total capital of JPY 7 million.

The level of investment among the shareholders is given in Table 4.1. The Izumisano PPS supplied around 2,000 MW in 2017, with the breakdown of electricity generation by local PV (17.26%), backup power supply (meaning procurement from Kansai Electric Co., Ltd in case of failing to secure its own electricity supply) (9.71%) and electricity trade through the Japan Power Exchange (JPEX) (73.04%) to achieve a high local renewable ratio compared with conventional electricity supplied by the existing utility (Figures 4.9 and 4.10).

Izumisano PPS total billing in 2017 was JPY 530 million, in which electricity sales counted for JPY 446 million (80%). After expenses, including those of electricity procurement (JPY 486 million) and administrative expenses (JPY 29 million), the net profit was JPY 18 million in 2017. The total accrued profit, from 2015 to 2017, was JPY 48 million. Figure 4.9 displays the relevant monetary flow regarding local energy development. The customers of Izumisano PPS are the public facilities within the city boundary, mainly owned and operated by the ISC itself, but there are some non-ISC customers (Suwa, 2020).

This case implies that the creation of a local power supplier, under the auspices of local government, can yield a new field of income and revenue created within a spatial boundary. The accumulated wealth may take the form of direct payment (through the supply of community energy, through employment, etc.), and indirect income via local tax revenues. Thus, the energy governance structure that incorporates local governments' power supplying institutions, though fierce competition with incumbent utilities, illustrates the possibility of using alternative business models to trigger institutional change and energy transition.

Discussion and conclusion

It is important to identify the dynamics and emergence of civil energy initiatives, and the motivation and impact of community and municipal energy ownership. The implication of the previously mentioned cases includes the types of benefits and scope of recipients of communal profit/benefits.

- Cooperative entities, which traditionally concentrated on retailing agricultural products, have evolved into "energy cooperatives" that produce and sell regional renewable electricity.
- There are increasing numbers of communities in Japan in which wider stakeholders are involved in developing community energy. The energy-related benefits are strongly connected with social and economic systems and the level of acceptance of renewable development. In the case of Waita village geothermal development, public acceptance was a barrier when the project was first envisaged by an entity that did not have sufficient local connection and dialogue. Motivation for geothermal development among community members emerged after they recognized the local problems, for example, decreasing population and the associated reduced local tax income and available budget for the community. The sense of urgency became the platform to urge residents to form a business entity and the relevant framework to carry

Table 4.1 Izumisano investment from stakeholders

	Share of total stocks
Izumisano City Government	83.3%
Power Sharing Co., Ltd	16.7%
Total	100%

Source: Based on Izumisanio PPS, re-created by author.

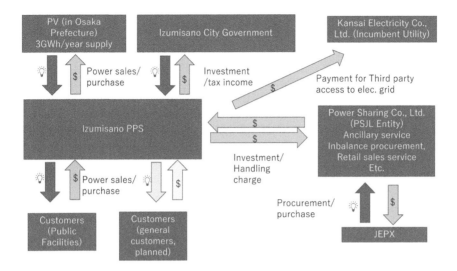

Figure 4.9 Izumisano PPS electricity and financial transactions.

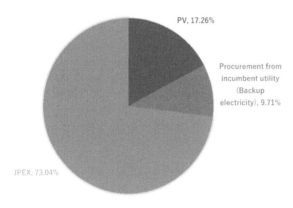

Figure 4.10 Origin of electricity supply (Izumisano PPS).

out the project, then to distribute benefits to local stakeholders. In those cases, the energy-related benefits are strongly connected to social and economic systems and the level of acceptance of renewable developments.

* The municipally owned multi-utility model, observed in many European countries, has become the iconic example for the creation of a governance layer, which has been studied and emulated by many localities in Japan. The observed formation of a community utility addresses a much more significant implication in countries where a centralized market structure long dominated (as in Japan), than in many of the European contexts where local government has a well-established role within the energy system and holds legal responsibilities in energy planning, addressing ownership and the operation of energy provision and services.

In the case of the cooperative model, members receive direct payment or dividends through the implementation of a given project. On the other hand, under the community business model, the beneficiaries of such a utility are the stakeholders of a business entity. The municipal energy utility model illustrates the scheme where residents receive indirect benefits as a means of municipal services. The services may include subsidized public transport and health and welfare provisions made available by the cumulative income through electricity and energy sales by civic utilities. The important departure from community energy production to civic utility schemes is that the spectrum of stakeholders is widened through the formation of civic utilities, although the benefits provided to each stakeholder may be less and indirect.

The widening of the stakeholder spectrum inevitably urges consideration on governance issues: relatively small and integrated governance as in the cooperative mode will no longer be effective in the civic utility approach because it involves a wider array of governance participants, ranging from municipal authorities, investors, community leaders and residents. The relations and interactions among stakeholders may become complex, where the potential conflicts among varied interests may need to be resolved. It is, therefore, vital to examine how governance interactions are emerging while solving the intensified imperative for agreeable communal benefit sharing.

To create an institution that can produce, manage and supply electricity at a given region in the form of a municipal civic utility in such a centralized (or quasi-centralized) market as the Japanese one is an important new venture. In Germany, the re-municipalization of utilities attracts wide attention (Creamer et al., 2018; Moss et al., 2015). In Japan, it is not "re"-municipalization, but "municipalization" itself of energy services that are witnessed. The political implication of such ownership of utilities is considered a route to regaining local engagement and a source for distributing financial and other communal resources, although it risks a rivalry with incumbent utilities (Figure 4.11).

Multiple lessons and recommendations can be drawn from our analysis of community benefit provision from local energy governance in Japan. By considering the overall development of local energy governance, it is not surprising that there is little deliberation on the latent interactions between the emergence of local energy governance and national government policy support. Surely, energy and electricity market liberalization has worked to enable the formation of new local energy organizations, however, there is also political negligence in optimizing the potential of new local energy schemes. National government involvement is minimum, whereas its support to the incumbent utilities largely remains in a more implicit manner, for example, through the newly created capacity market, which provides a payment based on the scale of power capacity to the incumbent utilities. In the national context, delivering local energy governance should be seen as the bottom-up drive to recuperate democratic decisions through energy-related choices.

Another dimension relates to the understanding of locality within democratic tenets. There is severe information asymmetry between the incumbent and local energy players in the energy market. Social and economic interests, and potential benefits from energy governance for local communities, are not properly framed in the media and associated communication. Therefore, the consumers of electricity do not have a chance of appreciating the values of the positive effects attached to their choice of electricity supplier. Current media approaches (e.g., TV and newspapers, as well as social media) to the public understanding of local energy governance do not allow innovative paths for fostering the democratic development of local energy governance and practices. Moreover, there is a colossal disparity in the volume of media advertising between the different sizes of electricity players. As a result, consumers are inclined not to realize the full optimization of local energy governance potential. There should be a more equitable approach to ensure the exposure of consumers to the choices of suppliers that can lead to avenues for rural energy development.

Further, the observed practices of community benefits with local energy governance developments in Japan can be tied to more innovative financial frameworks and products. By this, we mean that developers, either corporate or community-driven, should have further access to public facilities made available by a wider membership of society, with the use of innovative and emerging technologies. This can be done in a variety of ways, including the establishment and support from public funds, as the hometown investment fund addresses (Yoshino and Kaji, 2014) the use of an offset certificate to support renewable generation and other forms of emerging renewable financial products, such as green bonds, to secure finance for renewable development. In addition, the integration of economic and technological innovation, for example, nudging and blockchain solutions, may also increase the credibility of renewables, which, in turn, widens the opportunity to secure investment (Momsen and Stoerk, 2014). Further research should seek whether and how these financial opportunities benefit the community, which, in turn, exercise leverage to improve community benefit volume and its efficient allocation.

This analysis focuses on a case in Japan, where comparisons with other cases in different countries could be an interesting next step to understand the relation

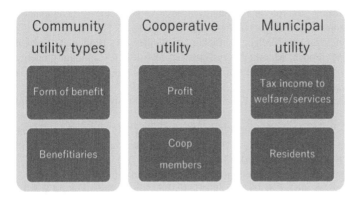

Figure 4.11 Initial typology of local energy arrangement.

between community ownership and geothermal development. It is even more important to see if the local energy governance framework can be applied in the wider international context. There is great potential for local economic growth in respective regions and communities and there is much promise for growth, but the returns may be further increased if community ownership is expanded. Community benefits may yet be modest and not large enough to influence the national trade and carbon accounts, but the implication of individual development can eventually impact on the cumulative benefits, and thereby affect the interests of public and private wealth decisions at large.

References

Allan, G., MacGregor, P. and Swales, K. (2011) 'The importance of revenue sharing for the local impacts of a renewable energy project – A social accounting matrix approach', *Regional Studies*, 45(9), pp. 1171–1186.

Atkinson, G. and Hamilton, K. (2020) 'Sustaining wealth: Simulating a sovereign wealth fund for the UK's oil and gas resources, past and future', *Energy Policy*, 139, Article 111273.

Bauwens, T., Gotchev, B. and Holstenkamp, L. (2017) 'What drives the development of community energy in Europe? The case of wind power cooperatives', *Energy Research and Social Science*, 13, pp. 136–147.

Bellaby, P. (2010) 'Theme 1: Concepts of trust and methods for investigating it', *Energy Policy*, 38(6), pp. 2615–2616.

Brisbois, M. (2020) 'Decentralised energy, decentralised accountability? Lessons on how to govern decentralised electricity transitions from multi-level natural resource governance', *Global Transition*, 2020(2), pp. 16–25.

Bulkeley, H., Broto, V.C. and Maasen, A. (2014) 'Low-carbon transitions and the reconfiguration of urban infrastructure', *Urban Studies*, 51(7), pp. 1471–1486.

Clausen, T. and David Rudolph, D. (2020) 'Renewable energy for sustainable rural development: Synergies and mismatches', *Energy Policy*, 138, Article 111289.

Cowell, R. (2007) 'Wind power and 'the planning problem': The experience of Wales', *European Environment*, 17(5), pp. 291–306. https://doi.org/10.1002/eet.464.

Cowell, R.J.W., Bristow, G.I. and Munday, M.C.R. (2011) 'Acceptance, acceptability and environmental justice: The role of community benefits in wind energy development', *Journal of Environmental Planning and Management*, 54(4), pp. 539–557.

Creamer, E., Eadson, W., Van Veelen, B., Pinker, A., Tingey, M., Braunholtz-Speight, T., Markatoni, M., Foden, M. and Lacey-Barnacle, M. (2018) 'Community energy: Entanglements of community, state and private sector', *Geography Compass*, 12(7), pp. 1–16.

Hamilton, K. and Hepburn, C. (2017) *National wealth*. Oxford University Press, Oxford, UK.

Hoppe, T., Graf, A., Warbroek, B., Lammers, I. and Lepping, I. (2015) 'Local governments supporting local energy initiatives: Lessons from the best practices of Saerbeck (Germany) and Lochem (The Netherlands)', *Sustainability*, 7(2), pp. 1900–1931.

Izumisano PPS, Homepage, https://izumisano-pps.or.jp/clause-documents/

Kikkawa, T. (2012) 'The history of Japan's electric power industry before world War II', *Hitotsubashi Journal of Commerce and Management*, 46, pp. 1–16.

Kimura, K. (2015) 'Grid parity – Solar PV has caught up with Japan's grid electricity', https://www.renewable-ei.org/en/column/column_20150730_02.php

METI (2013) 'Report of the electricity system reform expert subcommittee', https://www.meti.go.jp/english/policy/energy_environment/electricity_system_reform/pdf/201302Report_of_Expert_Subcommittee.pdf

Ministry of Environment Japan (MOEJ) (2017) 'White paper 2017', http://www.env.go.jp/en/wpaper/2017/index.html

Momsen, K. and Stoerk, T. (2014) 'From intention to action: Can nudges help consumers to choose renewable energy?', *Energy Policy*, 74, pp. 376–382.

Moss, T., Becker, S. and Naumann, M. (2015) 'Whose energy transition is it, anyway? Organisation and ownership of the Energiewende in villages, cities and regions', *Local Environment*, 20(12), pp. 1547–1563.

Nakayama, T., Raupach Sumiya, J. and Morotomi, T. (2016) 'Regional value added analysis of renewable energies in Japan: Verification and application of regional value added modelling in Japan', *Sustainability Research*, 6, pp. 101–115.

Nippon.com (2019) 'Japan relies on coal for 32% of electricity supply', https://www.nippon.com/en/japan-data/h00612/japan-relies-on-coal-for-32-of-electricity-supply.html

OECD (2012) *Inventory of estimated budgetary support and tax expenditures for fossil fuels 2013*. Paris: OECD.

Organization for Cross-regional Coordinator of Transmission Operators Japan (OCCTO) (2017) 'The status of changing suppliers', April 30, 2017, https://www.occto.or.jp/system/riyoujoukyou/170512_swsys_riyou.html

Rudolph, D., Haggett, C. and Aitken, M. (2014) 'Community benefits from offshore renewables: Good practice review', *ClimateXchange*. http://www.climatexchange.org.uk/files/7314/2226/8751/Full_Report_-_Community_Benefits_from_Offshore_Renewables_-_Good_Practice_Review.pdf

Schmid, B., Meister, T., Klagg, B. and Seidl, I. (2019) 'Energy cooperatives and municipalities in local energy governance arrangements in Switzerland and Germany', *The Journal of Environment and Development*. https://doi.org/10.1177/1070496519886013

Suwa, A. (2020) 'Renewable energy and regional value: Identifying value added of public power producer and suppliers in Japan', *Finance Research Letters*, 37, Article 101365.

Suwa, A. and Jupesta, J. (2012) 'Policy innovation for technology diffusion: Through a case with Japanese renewable energy public support programs', *Sustainability Science*, 7(2), pp. 185–197.

Suwa, A. and Sando, S. (2018) 'Geothermal energy and community sustainability', in *Proceedings of Grand Renewable Energy 2018 International Conference*, Yokohama, Japan.

Taghizadeh-Hesary, F. and Rasoulinezhad, E. (2016) 'Oil price fluctuations and oil consuming sectors: An empirical analysis of Japan', *Economics and Policy of Energy and the Environment*, 2016(2), pp. 33–51.

Taghizadeh-Hesary, F, Rasoulinezhad, E. and Kobayashi, Y. (2017) 'Oil price fluctuations and oil consuming sectors: An empirical analysis of Japan', *Economics and Policy of Energy and the Environment*, 2(19), pp. 33–35. https://doi.org/10.3280/EFE2016-002003

United Nations Framework Convention for Climate Change (2016) 'Paris agreement', https://unfccc.int/files/essential_background/convention/application/pdf/english_paris_agreement.pdf

Wade, J. Eyre, N. Hamilton, J. and Parag, Y. (2013) 'Local energy governance: Communities and energy efficiency policy', *ECEEE Summer Study Proceedings*, 3-031-13, pp. 637–648.

Walker, B.J., Wiersma, B. and Bailey, E. (2014) 'Community benefits, framing and the social acceptance of offshore wind farms: An experimental study in England', *Energy Research and Social Science*, 3, pp. 46–54.

Walker, G., Hunter, S., Devine-Wright, P., Evans, B. and Fay, H. (2007) 'Harnessing community energies: Explaining and evaluating community-based localism in renewable energy policy in the UK', *Global Environment Politics*, 7(2), pp. 64–82.

Walter, G. (2014) 'Determining the local acceptance of wind energy projects in Switzerland: The importance of general attitudes and project characteristics', *Energy Research and Social Science*, 4, pp. 78–88.

Warbroek, B., Hoppe, T., Bressers, H. and Coenen, F. (2019) 'Testing the social, organizational, and governance factors for success in local low carbon energy initiatives', *Energy Research and Social Science*, 58(101269). https://doi.org/10.1016/j.erss.2019.101269.

Yoshino, N. and Kaji, S. (2014) *Hometown investment trust funds: A stable way to supply risk capital*. Springer, Japan.

5 Barriers to renewable energy? A case analysis of the Garorim Bay tidal plant project in South Korea

Bomi Kim and Youhyun Lee

Energy transition is a critical policy goal in every country, and it is highly relevant to the UN SDG number seven – affordable and clean energy. In South Korea, there has been increasing interest in renewable energy with the growth in environmental problems. From 2008 to 2012, the Lee Myung-bak administration established various policies and institutions to increase the proportion of renewable energy in an attempt to ultimately achieve the national vision of green growth. For example, the renewable portfolio system (RPS) was adopted, which replaced the former feed-in tariff (FIT) system.

The new RPS regime forces major energy generation companies to implement certain renewable energy production quotas. The RPS system places varying levels of importance on each renewable source based on its rate of diffusion, ecological value and feasibility. Tidal power generation has become the biggest issue among renewable energy projects because tidal power has the highest weighting within the RPS regime.

In 2008, major Korean power companies began to plan for and establish various tidal power projects simultaneously, including Garorim Bay, Incheon Bay, Ganghwa Bay and Asan Bay. However, the most anticipated of these projects, the Garorim Bay tidal plant project, has caused frustration due to numerous controversies and conflicts that have arisen over the last five to six years. Moreover, it is very unlikely that a tidal power plant will be constructed in the future.

In this study, we attempt to clarify the factors that have obstructed this tidal power plant project, which has seemingly only caused a great deal of conflict and social cost. Regarding the interaction between institutions and policy actors as well as stakeholders, this controversial case has a crucial influence on follow-up tidal power plant plans and projects in other regions. This study also examines the problems involved in the process of establishing and executing renewable energy policies by analyzing these interactions between institutions and stakeholders.

DOI: 10.4324/9781003025962-8

The emergence of the local autonomy system in South Korea

After the 1953 armistice in the Korean War, all territorial and economic social systems had been destroyed. During the first half of the century, the main focus of the Korean government was overcoming national poverty. Due to the leadership of the Korean president, and the efforts of Koreans in overcoming poverty, Korea has surmounted its economic poverty. The fast economic growth in Korea was accomplished through the powerful leadership of its president, which provided a good excuse to implement central government-oriented policy from the 1950s through the 1980s.

In less than six decades, Korea has undergone a rapid political change. The framework and structure of Korea's political system were established in the Constitution of the Republic of Korea; the first Constitution was adopted on 17 July 1948. Since then, the Constitution has been amended nine times due to political upheaval related to the process of Korea's political development. For example, the current Constitution, which was amended on 29 October 1987, made substantial changes to the political configuration of Korea in response to the social demand for democratization. In this revision, the power of the president was reduced while the legislative power of the National Assembly was strengthened. The 1987 Constitution created an independent Constitutional Court that helped make Korea a more democratic and free society.

During the 20th century, three observations were made regarding the Korean system: the great stability of the administrative divisions, the lack of decentralization at the first level of the administrative organization and the rapid and strong urbanization of the country (the urban population accounts for 83% of the total population). These observations can help elucidate the history and characteristics of the Korean administration.

The Korean local autonomy system was adopted based on this political context. The legal basis for the current local autonomy system is Article 117 of the Constitution. According to Article 117 of the Korean Constitution, local governments must deal with administrative matters related to the well-being of the local population and manage the properties. They may adopt provisions on matters of local interest within the limits of subordinate laws and regulations.

With the implementation of the local autonomy system, the heads of local governments became interested in developing their regions and promoting the interests of their residents. This led to a change in the relationship and inter-influence between central and local governments, between inter-local governments, and between local governments and residents.

The discussion in this study regarding the Garorim Bay tidal power plant ranges from the centralized state system in 1977 to the beginning of the introduction of local autonomy in 1991, and the settlement of local autonomy in the region since the 2000s. It also concerns the period of post-settlement local autonomy, in which the power and participation of local residents increased, and the boundaries of the regions became clear. Different behavioral patterns emerged that depended on the inter-governmental level, the government–resident structure and power.

FIT to RPS: implementation of the RPS system

The renewable portfolio standard is a compulsory market-oriented policy instrument that was designed to promote the production of renewable energy. The Korean government already implemented a feed-in tariff system in 2001, but the following three turning points have since occurred (Lee, 2019).

First, the feed-in tariff system has not been able to effectively increase the production of new and renewable energy (NRE); the national NRE target was much higher than the actual achievement of the FIT system. During the Lee Myung-bak administration (from 2008 to 2012), the government was urged to promote an increased ratio of NRE in a short period with the national policy direction of green growth.

Second, there has come to be an excessive financial burden on the government. Since the FIT system guarantees a certain amount of profit over 20 years, this contract has aggravated the financial difficulties of the energy fund. While FIT and RPS share a common policy goal of promoting the supply of NRE, the government is not required to guarantee the long-term profit of power suppliers, eventually leading to less financial burden on the government (Lee and Seo, 2019). With these rationales, the Korean government decided to phase out FIT and pilot a test run of RPS, under the name of the Renewable Portfolio Agreement, with six state-owned power companies from 2009 to 2011 (Table 5.1).

The third point surrounds the preparation of the post-Kyoto regime, which was to be discussed in Paris in 2015. The possibility of becoming a target country for greenhouse gas (GHG) reduction obligations under the new climate change regime increased in 2012. The government needed to prepare for the new climate change regime, and the first task on this path was promoting NRE over fossil fuels.

After the FIT system was completely phased out in 2012, the RPS became the core policy instrument used to promote NRE in Korea. As a result, 13 power generation companies (now 26, as of 2019), including six state-owned companies, were designated at the time to implement certain NRE production quotas. These companies were set to be responsible for supplying 10% renewable energy by 2022, starting from 2% in 2012, and if they failed to fulfill their obligations, they would face certain legal penalties.

To fulfill the quotas, the obliged companies can independently construct renewable energy plants and directly acquire their own REC. They can also buy RECs from other companies. In the case of direct renewable energy generation, the company gets both REC and profit by selling any renewable energy generated. In this regard, buying REC is the preferred option for an obliged company.

Under the RPS system, the weighted value of REC differs from each renewable energy source in consideration of the effects of the environment, level of technological development, industrial effect, system marginal price, potentials of sources and GHG reduction, among others. As shown in Table 5.2, offshore wind power, tidal power and fuel cells have the highest weighted values among the various types of new and renewable energy sources. As tidal power has the highest weighted value of REC, many obliged power generation companies have planned

Table 5.1 Comparison of the RPS and FIT systems

Criteria	FIT	RPS
Instrumental mechanism	Voluntary and mixed	Compulsory
Domestic implementation	2001–2011	2012–present
System mechanism	Price manipulation by the government	Quantity manipulation by government
	Determination of quantity (market)	Determination of price (market)
Role of government	Guarantees adequate price	Guarantees adequate quantity
Market intervention	Direct intervention in pricing	Intervention in quantity
Compensation for cost	Utilization of electric power industry basis fund	Settled by KEPCO
Amount of compensation	Determined by market	Compensated according to the quantity determined by the government and the price set by the market
Unit price of compensation	Determined by government	Market makes primary decision, while the government and the price are set by the market
Import risk	No fluctuations	Double fluctuation (system marginal price [SMP] and renewable energy certificate [REC])
Project period	Applied to determined period (15 or 20 years)	Undetermined, depending on contract
Differentiation by technology	Differentiation among standard price by source	Compensated by weighted value

Source: Lee and Seo (2019).

for the construction of a tidal power plant. Regarding the scale of economy, the power generation companies have become eager to construct one mass power plant, rather than spreading out their efforts among several small-scale renewable facilities throughout the nation. From this standpoint, Garorim Bay was considered to be an ideal site for constructing one mass tidal energy plant.

Environmental impact assessment

The environmental impact assessment system is designed to propose the environment priority values in policymaking (O'Riordan and Sewell, 1981; Petts, 1999; Cashmore et al., 2008). The environmental impact assessment was first introduced in the United States in January 1970 under the National Environmental Policy Act (NEPA). It was introduced by the US Federal Government, starting

Table 5.2 Weighted values of REC for various NRE sources

Source	Weighted value	Types of energy and criteria	
		Types of installation	*Criteria*
Solar PV	0.7	Installations on general plot of land	Five nominated sites
	1.0		Over 30 kW
	1.2		Below 30 kW
	1.5	Installations on existing facilities (e.g., buildings)	
Other NRE	0.25	Integrated gasification combined cycle (IGCC), using by-product gas	
	0.5	Waste or landfill gas incineration	
	1.0	Hydro power, onshore wind, bio energy, RDF, waste gasification	
	1.5	Offshore wind power (less than 5 km of distance relay)	
	2.0	Tidal power, fuel cell, offshore wind (greater than 5 km of distance relay)	

Source: GGGI (2016).

with California, and was further introduced in Germany (1971), Japan (1972), Australia (1974), Thailand (1975) and France (1976). Studies on environmental impact assessments in each country suggest that sustainable development is the ultimate goal of environmental impact assessment (Lawrence, 1997; Gibson, 2000; Connelly and Smith, 2003; Doelle and Sinclair, 2006).

An environmental impact assessment has three main purposes. First, it serves as a policy tool to minimize the environmental impact of development and helps establish environmentally friendly plans by predicting environmental impacts by using various scientific tools. Second, it serves as a tool to help policymakers make decisions regarding policy directions (Cashmore et al., 2008; Heinma and Poder, 2010). Third, it is a means to minimize conflict in the process of consensus related to the environment among various stakeholders (Doelle and Sinclair, 2006; Cashmore et al., 2008; Isaksson, Richardson and Olsson, 2009; Gauthier, Simard and Waaub, 2011).

In Korea, the environmental impact assessment system was introduced in 1990 in an attempt to reduce environmental degradation caused by large-scale industrial development projects. Under the Environmental Impact Assessment Act, the Ministry of Environment has the right to consult on the environmental impact assessment of each development project. Before a project is approved, it must consult with the Ministry of Environment. If there is a specific reason to do so, the approval agency may consent even if the Ministry of Environment does not reach a compromise; however, due to the administrative burden, the approval agency generally supports each project based on the consultation of the Ministry of Environment. Therefore, the project approval is ultimately decided according to the consultation on the environmental impact assessment, and the actual decision-making authority and responsibility for the project is recognized by the citizens, stakeholders and the Ministry of Environment.

The environmental impact assessment is a procedural law, since the introduction of residents, discussions among all stakeholders and revisions to the plan are carried out through various stages. However, since it was introduced as a regulation for the first time, many people have recognized it as a regulatory law (Lee and Lee, 2010; Kim, 2004). Therefore, when conflicts arise regarding large-scale development projects, conflict unfolds according to each consultation procedure of the environmental impact assessment. The environmental impact assessment is also used as a winning strategy among stakeholders.

When conducting a large-scale development plan or project, the environmental impact assessment is almost the only procedure that allows residents and others to receive information about the project or to express their opinions. The environmental impact assessment system consists of the strategic environmental impact assessment (SEA) and the environmental impact assessment (EIA), each of which includes the same procedures for local participation. There are two ways in which residents or NGOs can participate in the environmental impact assessment system: presenting opinions through residents' briefing sessions or participating directly as members of the environmental impact assessment committee.

Case record

Data collection

The data were collected through interviews and newspaper articles. We interviewed government officials in the Ministry of Environment and the Ministry of Trade, Industry and Energy; these officials represented the governmental side. We also conducted interviews with Korea Western Power Co., Ltd, which is responsible for implementing RPS duties. Further, we interviewed EIA implementing parties, local residents, an NGO (Environmental Action Union), the Korea Environment Institute and other experts who participated in the EIA process.

Newspaper data were gathered with a focus on three different categories: the (1) conservative stance, (2) proactive stance and (3) regional stance. Articles from Joong-ang, Chosun and Donga made up the conservative stance. Hangyorae, Gyunghyang and Seoul made up the proactive stance. Finally, articles from Choongchung Today, Joongbumaeil, Joongbu, Daejeon and Joongdo made up the regional stance.

Scope of research

The periodical scope covers the period from January 1977 through October 2014. The construction of a tidal power plant in Garorim Bay first began to be discussed in 1977; however, after conducting several economic feasibility studies, the project did not continue due to its perceived low feasibility. Later, the project was delayed because the environmental impact assessment process did not reach a consensus. Finally, as it had not made any progress, the project was terminated. The contractor did not abandon the project by its own decision, but rather the deadline for the shared reclamation plan expired during the EIA process due to its slow pace. The valid period for a shared reclamation plan is five years; if the validation expires after five years, the SEA consultation process must be re-started. Considering these facts and the current situation, the project is assumed to have ended.

The spatial scope mainly includes the coast of Garorim Bay in Seosan, Choongchunnamdo. The coast is located in northern Taean, Choongcheongnamdo. This coast is surrounded to the north by Taean Peninsula and Cheonsu Bay, to the south by Taean-eup and to the west by Wonbukmyon and Iwonmyon; Palbongmyeon, Jigok-myeon and Daesan-myeon are located in the eastern portion of this spatial scope. This gourd bottle-shaped bay has a semi-closed entrance. Compared with a narrow entrance, the bay has a large space inside, making it an ideal topographical condition for a tidal power plant site.

Timeline

The first construction plan was developed by the Garorim Tidal Power Plant Consortium, which was founded by the Korea Western Power company and three construction firms: Lotte, Posco and Daewoo. The construction period was approximately 83 months, and the total cost was one trillion Korean won. The plant was designed to adopt single flow type generation (at low tide), and its estimated yearly generation was 950 GWh, which would be equivalent to 40% of the Choonchungnamdo region's annual total power demand (for household). The estimated installed capacity was 530 MW, with 20 generation organs (26 MW per organ) (Table 5.3).

Stakeholder attitudes and relationships according to the implementation of EIA

BEFORE IMPLEMENTING THE EIA

In 1977 and 1995, feasibility studies were conducted. However, due to the low economic feasibility of KEPCO, the project was not ultimately undertaken. In 1995, KEPCO conducted a detailed review with a strong commitment to the Ministry of Trade, Industry and Energy.

- **Agreeing parties: central – local (Chungcheongnam-do)**

Since the 1970s, the central government has maintained a strong will to establish a Garorim tidal power plant, and most of the related ministries agreed with the plan until residents began voicing their opposition. The Ministry of Trade, Industry and Energy took the initiative regarding the project in an effort to develop alternative energy. In 1995, the company also conducted its own feasibility review, included it in the long-term electricity supply and demand plan (1995) and confirmed the KEPCO and construction plans. Since then, the Ministry of Commerce, Industry

Table 5.3 Chronology

Date	Context	
7 April 2008	Request for Prior Environmental Review Committee (Western Power Co. → Ministry of Land and Sea)	Scoping of strategic environmental impact assessment
16 October 2008 and 18 December 2008	1st and 2nd Prior Environmental Review Committees (Ministry of Land and Sea)	
20 January 2009	Prior Environmental Review Committee (Draft) handed over opinion (Ministry of Land and Sea)	
2 March 2009	Prior Environmental Committee (draft) submitted (Ministry of Land and Sea)	Strategic environmental impact assessment
27 April 2009	Prior Environmental Committee (draft) briefing session (one meeting in Taean, one meeting in Seosan)	
30 June 2009	Inscription of the original bill; supplementation of first, second and third bills	
28 October 2009	Consultation on the original bill of Prior Environmental Committee	
18 November 2009	Announcement regarding change of the Basic Plan for Shared Water Reclamation (n. 2009-1068)	
June 2010	Submission of evaluation plan and request for deliberation (Garorim Bay tidal power → Ministry of Knowledge Economy)	Scoping of environmental impact assessment
8 July 2010	Composition and discussion of Deliberation Committee on Environmental Impact Assessment Plan (Ministry of Knowledge Economy)	
6 October 2010	Review of evaluation plan (Ministry of Knowledge Economy → Garorim tidal power)	
11 October 2010	Submission of EIA evaluation paper (draft) (Garorim tidal power → Taean local authority)	Environmental impact assessment
27–29 October 2010	Resident briefing sessions on EIA evaluation paper (draft) in Taean and Seosan	
23 December 2010	Conversation session in Choongnam	
23 February 2011	Resident briefing session	
10 February–25 March 2010	Submission of required supplementary action plan of Seosan-si (one to two times)	
12 April 2011	Public hearing session on EIA paper	
13 April 2011	Submission regarding notice of result (Taean and Seosan)	
May 2011	After submitting the Environmental Impact Assessment paper (draft), submission of the first, second, third, fourth and fifth revised papers	
October 2014	Rejection of documents by the Ministry of Environment; expired deadline for Shared Water Reclamation landfill master plan	

Source: Modified by authors based on the Report by Korean Western Power Co. (2007 to 2014).

and Energy included the Garorim Bay tidal power plant in the establishment of the 2nd National Energy Basic Plan (2002–2011). In addition, the Ministry of Maritime Affairs and Fisheries included the Garorim Bay tidal power plant in its 2004 plan.

During this period, local governments were either in favor of or neutral toward the plan and awaiting more specifics, at which point the residents began voicing their opposition; until the early 2000s, residents did not oppose or consider the government's plan. Accordingly, the development projects were conducted in accordance with the central government opinion or according to the local development logic. Indeed, in 1991, the Governor of Chungcheongnam-do requested that the central government construct a power plant for local development at the National Audit. In 1995, Chungcheong province announced the plan for a Garorim tidal power plant in the Chungcheong region development plan. In these contexts, the local and central governments shared a common positive standpoint on the project.

Another local government, Taean, which was in favor of the position, was also an active participant in the discussion processes of these construction plans. In 2006, the Mayor of Taean demonstrated his willingness to support the construction plan in a municipal speech.

- **Opposing parties: central – local (Seosan)**

Seosan, which has consistently been opposed to such a plant from the outset, submitted proposals against the tidal power plant to the ministries involved in 2005, as the project was being specifically discussed and plans were being specified. In 2005, Seosan submitted a recommendation to the Ministry of Maritime Affairs and Fisheries not to include a plan for tidal power generation in the Garorim Bay Management Area Plan (which was not an official procedure). In 2007, the city tried to establish a comprehensive strategy at the city level by gathering opinions from local residents and environmental groups.

- **Neutral parties: local residents and companies**

In the initial stage of the project, residents and NGOs were aware of discussions regarding the project, but they did not know the details. However, since 2005, as detailed discussions have come to be held, some concerns have begun to emerge for those involved in the fisheries in Garorim. In the 1990s, the construction company accepted the opinion of the Ministry of Commerce, Industry and Energy, but they thought that the project was not sufficiently feasible. Thus, in the early days, they were not interested in the project and did not express any particular position.

STAGE OF IMPLEMENTING THE SEA

Previously, environmental and traffic impact assessment procedures were mandatory, but the SEA was strengthened with the revision of the Framework Act on Environmental Policy in 2006. In addition, since conflicts were incurred gradually, the Ministry of Environment rejected the existing environmental and traffic impact assessment procedures and established a new SEA procedure in 2008. The SEA consultation was completed on 28 October 2009. On 18 November 2009, the Ministry of Land, Transport and Maritime Affairs revised and announced the Basic Plan for Shared Landfill, which approved and revealed the change to the purpose of the land.

- **Agreeing parties: central government (Ministry of Energy and Ministry of Land)**

The Ministry of Knowledge Economy and Ministry of Land, Transport and Maritime Affairs, which are parts of the central government, were in favor of the plan but shifted responsibility and authority toward the Ministry of Environment as conflicts around the environmental impact assessment system began to spread.

The approval authority of the SEA was the Ministry of Land, Transport and Maritime Affairs. In total, ten members participated in the SEA committee: the Korea Environment Institute; the Ministry for Food, Agriculture, Forestry and Fisheries; the Taean local government; the Seosan local government; and NGOs. Only one NGO actually opposed the plan and rejected the request by dissidents who asked for its attendance.

- **Opposing parties: Seosan, local residents**

As Seosan city's opposition was intense, its opinions were communicated to each related organization. Other important opposing parties were NGOs and residents. NGOs may not hold residents' demonstrations through armed demonstrations, publicize them through performances and media, publicly issue press statements or submit appeals to government departments and the Blue House. Instead, they promote the involvement of politicians who value the public importance of residents. In fact, as the opposition activities of citizens and civic groups became more intense, the media took increasing notice. Since then, interest in Garorim tidal power has grown nationwide.

- **Neutral parties: Ministry of Environment, Chungcheongnam-do**

The Ministry of Environment expresses neutrality and feels burdened with opinions regarding the project. Therefore, the focus was on delivering opinions and centered on the views of experts or the Korea Environment Institute for each procedure.

> Because it was the first large-scale tidal power generation, it was difficult to make any decision (…) it was difficult to predict what would happen. (Interview with Ministry of Environment, 12 November 2015)

Chungcheongnam-do began to take a neutral position by changing its position in favor of Seosan city's initial opposition. The governor of Chungnam province insisted that the governor's authority is outside the right of consultation, stating that he would like to examine whether the majority of residents' consent, empirical data on environmental changes and procedural legitimacy had been secured.

STAGE OF IMPLEMENTING THE EIA

After the SEA was negotiated and agreed upon, the project would be implemented with the approval of the Ministry of Knowledge Economy (hereinafter referred to as the Ministry of Trade, Industry and Energy), which is the approval agency for the project. Therefore, the consultation procedure regarding the EIA was the last opportunity to block the project. From the perspective of the construction company, agreement to the SEA would be recognized as approval of the project.

- **Agreeing parties: Ministry of Trade, Industry and Energy and the operator (construction company)**

The Ministry of Knowledge Economy continued to specify details of the Garorim project in the 4th Basic Power Supply and Demand Plan. The ministry focused on the development of new and renewable energy sources. In the SEA, the Ministry of Energy had no direct authority, so it could only conduct meetings or send cooperation documents. However, in the EIA stage, the ministry was able to participate as a direct approver, and it had the authority to form the Environmental Impact Assessment Council.

At the beginning of the EIA process, the operator did not consider the population. However, as the supplementary demands of local residents to resolve conflicts were repeated, the Ministry of Environment became interested in conflict resolution. After all, the conflict with the opposing side was so deep that it could not be resolved.

> One of the most regrettable things was that I didn't listen to the people in the early days. It would not have happened if we were interested in listening to what we could not solve, such as compensation outside of law, and trying to solve it. (Interview with the operator, 19 November 2015)

- **Opposing parties: Chungcheongnam-do, residents and NGOs**

As the conflict escalated, Chungcheongnam-do took the opposite position and assumed an active role. In February 2014, the Ministry of Environment responded positively to the Chungcheongnam-do request for a review. A review committee, composed of 16 private experts, was formed to review the environmental impact assessment for two months. A faithful environmental impact assessment should be based on the results of a thorough on-site survey that allows for sufficient time and should be prepared by predicting the prospect of change in accordance with the project implementation with the goal of establishing a plan to reduce environmental damage (15 April, 2014, Chungcheong Today).

Residents and NGOs were the key actors among the opposing parties at the EIA level, similar to the former stage of SEA. After the briefing session, officials were no longer authorized; therefore, they visited government agencies, politicians and other relevant administrative agencies to make their anti-construction appeals. The issue of the Garorim Bay tidal power project was raised through various routes, including the signing movement. In particular, in June 2011, the original draft of the Environmental Impact Assessment was submitted to the Ministry of Environment, and the civil environmental organizations drew up the Seosan-Taean Solidarity Conference for the Whitening of Garorim Bay Dam. In this conference, the Seosan-si Fishing Village Council, the Taean Fisheries Village Council, the Korea Fisheries Federation and 33 organizations in Seosan and Taean – including the Seosan Taean Federation, Taean Association, Seosan Taean Environmental Movement Association and Seosan Eastern Mayor Association of Mayors – participated.

> Resolve the conflict later (…) These opinions keep coming out, so I did a lot to act.
>
> We made several proposals to the operator. Let's do it together, but they ignored [us]. (Interview with an NGO, 21 November 2014)

- **Neutral parties: Ministry of Environment**

At this time, the opposition's activities increased, and the press was under pressure. In addition to the Ministry of Energy, the relevant institutions in Chungcheongnam-do, Seosan-si and Taean placed pressure on the Ministry of Environment. Due to difficulties associated with forecasting issues related to large-scale projects, the burden was higher than expected. Therefore, the Ministry of Environment could not make any decision and showed an indecisive attitude. By 2014, the Ministry of Environment only provided the opinions of the respondents based on three supplements, and no pros or cons were given. In the end, as the period for SEA consultations has expired, they are by default considered as having evaded liability until they had no choice but to give up.

> There are three criteria for evaluating the assessment: clear environmental impacts, predictions of their impacts, and specific concrete measures for their impacts. But since oceans tend to change so much, they have no choice but to predict the impact or reduce the impact. That was the biggest problem. (Interview with the Ministry of Environment, 12 November 2015)

Discussion

Interaction of institutions and actors

Institutional change affects the behaviors of stakeholders who are involved in the institution. As the institution changes, the stakeholders seek to maximize their benefit by thinking about the transaction cost, deciding on their own behaviors and interacting with other stakeholders. In the Garorim case, RPS and EIA are the institutions that have a crucial influence on stakeholders.

Before the RPS system was implemented, the Garorim Bay area had been discussed as a location for tidal power projects since the 1970s for the purpose of local development. In 1995, a concrete construction plan was established, and an implementation agency was selected, but the economic feasibility of the plan was ultimately questioned. However, with the introduction of RPS, tidal power generation has become the most feasible project. As a result, in addition to the Garorim Bay area, several tidal power plant projects were simultaneously discussed in Ganghwa Bay and Incheon Bay.

EIA was introduced in the 1990s, and it is one of the continuously changing institutions. The biggest change in the EIA system is that the procedures have been subdivided and the number of participants has grown. While limited, the opportunities for citizens and NGOs to participate in the large development project also increased, as did the influence of their input. EIA was developed as a regulatory or scientific and technological process, but it soon turned into a planning process system in which the plan could be revised through consultation with stakeholders through numerous stages and procedures.

However, complex interactions have been observed among the group that regards EIA as a regulatory or scientific and technological process (central government, operators) and the group that considers it a participatory process (experts, local governments), as well as the other group, according to their interests and values. Whether they agree on the plan or not, these interactions affect the position and strategy of each actor, according to each procedure of EIA.

In the end, the distrust and conflict between the various parties involved aggravated the situation. The EIA's consultation authority put off the final decision and let the operator abandon the project, as the consultation's validation date set in the SEA stage passed. The operator's investment in terms of only the environmental impact assessment was 7.9 billion won.

In the end, the EIA, which aimed to improve the process of sharing opinions, fueled conflict and encouraged distrust of the different actors, causing very high social costs, due to the different understandings of the participants' institutions and their pursuits of their own benefits.

Inter-governmental conflict: central vs. local, local vs. local

When planning and establishing local energy policies or institutions, conflicts between central vs. local governments and local vs. local governments, and, in turn, cooperation, are becoming increasingly important. Prior to the implementation of the local autonomy system, the central government established policies in a top-down manner, meaning that local governments were only responsible for implementing policies. Despite feelings of dissatisfaction among several local governments, severe conflict was rare at that time. However, independence and competition among local governments have intensified with the implementation of a local autonomy system. Decentralization of public administration has led to increasing cases of both conflict and cooperation with local and central governments or other neighboring local governments.

This phenomenon was also evident in the Garorim Bay tidal power project. Before the implementation of the local autonomy system, local governments, such as Chungcheongnam-do, Seosan and Taean had to make requests to the central government regarding the establishment of plans or withdrawal of projects. However, after the local autonomy system was implemented, the local councils came to be influenced by local elections. Local councils began expressing their strong desires to the Blue House (president) and central government (for their regional benefit), as well as expressing a stronger willingness to meet local demands.

In addition, conflicts between local governments began to intensify, specifically between Seosan and Taean. As a result, Chungcheongnam-do, a metropolitan local government between Seosan and Taean, attempted to set up a conflict-related ordinance to resolve conflicts but failed to achieve any solutions.

Concluding remarks

This study highlights several factors that caused the failure of the Garorim tidal power plant project. At a fundamental level, the government overlooked the renewable energy business, and subsequently, the factors that led to most of the social costs were as follows. First, the new local autonomy system was an important factor. Until the 2000s, the central government had controlled the area, and local governments rarely resisted the central government's initiatives. However, with the implementation of the local autonomy system and the election of local government heads, the heads of the local governments and provincial councils began to act as local representatives. The Ministry of Trade, Industry and Energy (MOTIE), who attempted to carry out the project, did not make any efforts to involve the local governments and residents. This led to strong opposition from local residents, and this opposition was then supported by local governments; this ultimately became a major obstacle to the project.

Second, the RPS system was a powerful regulation established by the central government to strengthen the use and production of renewable energy. It was necessary to introduce the RPS system to drive rapid growth in renewable energy. However, in setting the policy direction for renewable energy in South Korea, the weight of the RPS system was not properly determined in consideration of the circumstances in South Korea. Since the announcement of the RPS system, most of the major power generation companies have pursued the highest weighted tidal power projects, and these companies have determined that tidal power plants are the most economical solution for this purpose.

However, tidal power generation projects are not suitable for South Korea's natural environment. In addition, most of the areas in which tidal power projects have been considered are in the West Sea, but tidal-flat and foreshore fishing are common in this area. Because of this, severe opposition was anticipated from the start. Local residents and environmental NGOs were anxious that the seawater would be contaminated if tidal power plants were built. The failure to take into account the various circumstances that may occur in the implementation process made the RPS system hinder, rather than support, the development of renewable energy.

Finally, the policy actors in the EIA process were another major factor. The renewable energy project was closely related to the EIA system, which requires various stakeholders to participate in each procedure. The EIA has changed to promote the participation of residents to prevent conflict, but the MOTIE did not consider dissent, such as dissent from residents. Instead, they considered the EIA as a technological, formal procedure, and they wanted to proceed even if objections were raised. However, the dissenters used various means to reinforce the logic of their opposition, and they used the EIA to apply political pressure on the Ministry of Environment, who was the final arbiter. Institutions that are designed to involve multiple stakeholders are not created by individuals or individual groups. The Garorim Bay tidal power project was carried out without understanding the characteristics of the EIA, and by failing to recognize the necessity of involving multiple stakeholders, it was set up to fail from the beginning.

References

Cashmore, M., Bond, A. and Cobb, D. (2008) 'The role and functioning of environmental assessment: Theoretical reflections upon an empirical investigation of causation', *Journal of Environmental Management*, 88(4), pp. 1233–1248.

Connelly, J., Smith, G., Benson, D. and Saunders, C. (2003) *Politics and the environment: From theory to practice*. London: Routledge.

Doelle, M. and Sinclair, A.J. (2006) 'Time for a new approach to public participation in EA: Promoting cooperation and consensus for sustainability', *Environmental Impact Assessment Review*, 26(2), pp. 185–205.

Gauthier, M., Simard, L. and Waaub, J.P. (2011) 'Public participation in strategic environmental assessment (SEA): Critical review and the Quebec (Canada) approach', *Environmental Impact Assessment Review*, 31(1), pp. 48–60.

GGGI (2016) *Korea's green growth experience: Lessons learned and outcomes*. Seoul: GGGI Report.

Gibson, R. (2000) 'Favouring the higher test: Contribution to sustainability as the central criterion for reviews and decisions under the Canadian Environmental Assessment Act', *Journal of Environmental Law and Practice*, 10(1), pp. 39–54.

Heinma, K. and Põder, T. (2010) 'Effectiveness of environmental impact assessment system in Estonia', *Environmental Impact Assessment Review*, 30(4), pp. 272–277.

Isaksson, K., Richardson, T. and Olsson, K. (2009) 'From consultation to deliberation? Tracing deliberative norms in EIA frameworks in Swedish roads planning', *Environmental Impact Assessment Review*, 29(5), pp. 295–304.

Kim, W.-J. (2004) 'A study on public law regulation and remedies for environmental infringement', *Environmental Law Review*, 26(1), pp. 65–84.

Korea Western Power Co. (2007) *EIA Report on Garorim bay tidal power plant* (Unpublished)

Korea Western Power Co. (2010) *EIA Report on Garorim bay tidal power plant.* (Unpublished)

Korea Western Power Co. (2011) *EIA Report on Garorim bay tidal power plant.* (Unpublished)

Korea Western Power Co. (2014) *EIA Report on Garorim bay tidal power plant.* (Unpublished)

Lawrence, D.P. (1997) 'PROFILE: Integrating sustainability and environmental impact assessment', *Environmental Management*, 21(1), pp. 23–42.

Lee, W.-h. and Lee, H.-y. (2010) 'Changes in Korean environmental regulations and their characteristics', *Korean Policy Sciences Review*, 14(3), pp. 29–54.

Lee, J.S. (2019) 'Green growth in South Korea', in *Handbook on green growth.* Chaltenham Edward Elgar Publishing.

Lee, Y. and Seo, I. (2019) 'Sustainability of a policy instrument: Rethinking the renewable portfolio standard in South Korea', *Sustainability*, 11(11), p. 3082.

O'Riordan, T. and Sewell, W.D. (1981) *Project Appraisal and Policy.* Chichester: John Wiley, pp. 297–302.

Petts, J. (1999) 'Public participation and environmental impact assessment', in *Handbook of environmental impact assessment*, Blackwell science: Oxford. vol. 1, pp. 145–177.

Part III

Local partnerships for the development of renewable energy at the local level: citizens, communities and companies

6 Local public companies, local authority shareholders and electricity

Rarely one, never two, always three

Marie-Anne Vanneaux

Local public companies and electricity have close ties that are part of a common and fertile history.[1] Indeed, the French electricity sector was built locally and nationally and then gradually liberalized. It currently comprises four activities: production, which can be carried out by private companies in addition to the incumbent operator *Electricité de France* (EDF); transmission, which uses a network monopolized by *Réseau de Transport de l'Electricité* (RTE); distribution, which brings medium and low voltage electricity from the transmission network to the point of delivery to customers and is carried out by two network operators, which also enjoy a legal monopoly, *Enedis* and local distribution companies (LDCs); and finally, the supply of electricity. Considered as a commodity, electricity is sold by EDF or the LDCs at regulated tariffs or offered freely (market offer) to customers who have asserted their eligibility.[2] Over the last ten years, in the context of growing awareness of climate issues and the need to use renewable energy (RE)[3] from now on, local authorities have seen their legitimacy to be involved in these areas considerably increased. For example, Article L 100-2 of the Energy Code requires them to act in the area of energy demand management, encouraging them to promote energy efficiency and *sobriété* (which could be translated as energy conservation) to diversify supply sources and increase the share of renewable energy in the final energy consumption. The laws of 3 August 2009 and 12 July 2010,[4] known as the Grenelle I and II laws, thus give new functions to local authorities in the field of renewable energy, which is, in fact, easier to develop at that scale because the resources are local, and therefore decentralized, and because local authorities have a long experience in the supply and production of electricity (Daydié, 2016, p.34 and seq. p.239 and seq.; Marcou, 2007, p.20). Indeed, municipalities and their groupings have historically been the distribution organizing authorities.[5]

The territorialization of the national sustainable energy policy, with its environmental imperatives, therefore, imposes new responsibilities on local authorities and at the same time gives them the opportunity to (re)position themselves at the center of electricity policy. In this context, local public enterprises (LPEs) prove to be legal tools particularly well adapted to the concrete implementation of the responsibilities granted to local authorities in this field.

DOI: 10.4324/9781003025962-10

Not defined by law, these companies, sometimes also called local public operators (Bizet, 2012, p.15; Nicinski, 2019), actually cover several types of structures, which have joint-stock companies[6] in common, and are held in a majority or preponderant way by the local authorities and their groupings and whose function is to take charge of an activity of general interest. The family of LPEs thus includes the most numerous and oldest semi-public companies (SPCs), local public companies (LPCs) and single-purpose semi-public companies (SPSPCs).[7] The intervention of LPEs in the field of electricity actually dates back to the beginning of the 20th century, when the law of 15 June 1906 on energy distribution qualified access to electricity as a "public service" and granted municipalities and intermunicipal syndicates by delegation, ownership of electricity distribution networks as well as the status of licensing authorities or organizers of public energy distribution.[8] Until the beginning of the 1930s, "a mosaic of independent power stations and local networks operated within the framework of municipal or inter-municipal companies or concessions [taken over by SPCs]" (Belot and Juilhard, 2006, p.39) and agricultural collective interest companies (ACIC) developed throughout the country.

The first major transformation in the electricity sector occurred with the law of 8 April 1946. This law not only nationalized the production, transmission, import, export and distribution of electricity[9] but also created *EDF-GDF (Electricité de France-Gaz de France)*, a public industrial and commercial institution. EDF-GDF, now a national public company, was entrusted with all the activities listed earlier without facing any competition since the law had granted it both a monopoly and the status of a mandatory concessionaire. Consequently, even though Article 36(1) of the Nationalization Law reaffirms that local authorities remain the owners of their installations and that concession remains the main method of managing the public electricity distribution service, they can in principle no longer choose partners. However, by way of an exception, Article 23 of the law of 8 April 1946 allows local authorities to keep their network and existing electricity supply operators, which are now known as non-nationalized distributors (NND).[10]

At the end of nationalization, the electricity sector had 360 companies, consisting of 100% public corporations (*régies*), SPCs and ACICs (Gabillet, 2015, p.124; Allemand, 2007, p.31),[11] the first entities (public corporations) being the biggest share. The low representation of the SPCs remained the case for a long time. In sum, since 1949,[12] it has been forbidden to create new NNDs of any legal form. In addition, the financial difficulties that these operators have always had to face because they operate in a segment of the electricity market that is not very profitable, and which requires heavy investment, encouraged them to group together (Gabillet, 2015),[13] and only eight LPEs were established between 1949 and 2000.[14] This trend of the energy sector differs from what has been experienced at the same time by the local public sector as a whole, where the number of SPCs increased from 500 in 1964 to 1199 in 1999 and 1306 in 2000 (Da Rold, 2008, p.181).[15]

The beginning of the 2000s was a pivotal period in the involvement of local authorities not only in local electricity distribution but also in production. The law of 10 February 2000 on "the modernization of production and the development of the public electricity service"[16] maintained the monopoly of EDF and the LDCs on electricity transmission and distribution networks and reaffirmed that local authorities remained the sole owners and licensors of these networks. These two sectors are therefore not liberalized and the *status quo ante* remains for these segments. On the other hand, the activities of generating and supplying electricity to eligible customers are open to competition. This trend is strengthened with the laws of 3 January 2003, 4 August 2004 and 7 December 2006[17] prohibiting integrated companies from aggregating their electricity supply and distribution activities. The LDCs serving more than 10,000 customers must therefore drop supply unless they place them in an entity legally separate from them.[18] They must also henceforth allow alternative suppliers non-discriminatory access to their network unless the electricity is supplied to non-eligible customers, in which case the LDCs still have exclusive rights. However, the law of 10 February 2000 goes even further and is not limited to redefining the role of traditional electricity distribution operators. It triggers an intensification of the involvement of local authorities in the field of electricity production. Municipalities and their groupings can now develop and operate any new hydroelectric installation with a maximum power of 8000 kW, any new installation using other renewable energies, any new installation for the energy recovery of household waste or any new installation for supplying a heating network (Article L 2224-32 Local Governments Code (LGC)). The new Article 2224-33 of the LGC adds that, in the context of public electricity distribution, the authorities awarding concessions for electricity distribution may develop, operate directly or have operated by their electricity distribution concession holder any local electricity production installation where such an installation is such as to avoid, under good economic conditions, in terms of quality, safety and security of the electricity supply, the extension or reinforcement of the public electricity distribution networks under their jurisdiction. By mentioning twice the possibility for local authorities to produce electricity from renewable energy sources and by adding to them, through Article 2224-34 of the LGC, the power to carry out or support energy management measures, the law of 10 February 2000 does not simply increase their responsibility in terms of energy production, it "marks a real starting point in [their] consideration of sustainable development concerns" (Belot, 2013, p.19).[19] This involvement of local authorities was further strengthened when, five years later, the Programme Law of 13 July 2005[20] "setting out the guidelines for energy policy" authorized the public operators in which local authorities are shareholders to produce electricity from wind farms, hydroelectric plants or photovoltaic panels.

The two so-called "Grenelle laws" consolidate the possibility for local authorities to become directly involved in the production and promotion of renewable energy, while the first, dated 3 August 2009,[21] only indirectly affects the legal regime for the distribution and production of electricity, the law of 12 July 2010[22] increases the powers of local authorities in terms of network energy management and now gives *départements* and *régions* the capacity to develop, or have developed, operate or have operated, production facilities using renewable energies.[23]

As for the Energy Transition Law for Green Growth of 17 August 2015,[24] it provides an ambitious overhaul of the French energy system, which must now turn promote renewable energy and encourage the reduction of greenhouse gas emissions and the reduction of fossil fuel consumption. In order to achieve these objectives, local authorities are once again being called upon to contribute, first in the traditional way by using their usual operational relay, the LPEs, which since 2010, in addition to SPCs, include LPCs[25] and SPSPCs.[26] In addition, in a more original way by being authorized for the first time to become non-majority shareholders in companies working in the field of energy production and management, in particular, renewable energy.[27]

The French electricity ecosystem has therefore always depended, at least in part, on local authorities and their local public companies. However, gradually, local authorities have indeed acquired greater freedom in the choice of their investment tools, especially in the field of electricity production. Initially limited to SPCs and then to LPCs, this freedom can now be expressed for the first time through minority public shareholders. Local shareholding in the electricity sector is therefore now multifaceted and new prospects for local intervention may emerge.

Local public electricity companies with majority public shareholdings: the most common legal form

Local and regional authorities have a range of local public enterprises that they can use to implement local energy policies, particularly renewable energy policies. Among these, SPCs and LPCs have in common that they are controlled by their shareholder authorities, have an identical corporate purpose and are the most widely used structures. They represent 66% of LPEs operating in the energy sector,[28] bearing in mind that 72 SPCs and 5 LPCs[29] currently make up the more specific local public sector of electricity distribution and production. Although their number has continued to increase since the laws of 10 February and 12 July 2010,[30] it should, however, be noted that only nine NNDs exist in this form[31] and that at the same time no LPCs carry out distribution activities, whereas the NOME Law[32] created this new option for local authorities.

The legal formula adhered to by the majority of local authorities is, therefore, that of the SPCs, while the LPCs are not favored by them.

The semi-public company, an appropriate legal formula favored by local authorities

The visible success of SPCs can be explained by the fact that, as a limited company whose shareholding must be between 50% and 85%,[33] it can be created by a simple decision of the local authorities. It is flexible and can be used as a subcontractor to provide services for third parties. Its scope of activities is broad,[34] which enables it not only to produce electricity or carry out energy management activities but also to extend these activities materially and geographically if the achievement of its corporate purpose so requires. The SPCs may thus diversify and create subsidiaries provided,[35] however, that the local authorities that are shareholders expressly authorize them to do so, that the associated activities are complementary to the corporate purpose and that they correspond to the powers of the public shareholders (Brameret 2011). This structure can thus adapt to any kind of local energy project and intervene at each stage.

The economic or even financial advantage that this tool provides is minor for local authorities. The SPC is in fact a way to ensure that the renewable energy production activity of local authorities is consistent with their "local development objectives. [It allows them] to maximize the economic returns of these investments for the sole benefit of the territory and its inhabitants" (Jourdan, 2012, p.3). As shareholders, the local authorities benefit from the potential financial health of the SPC, which results in particular from the resale of electricity within the framework of the purchase obligation.[36] The public shareholders can then either share the dividends or reinvest them in production activity, energy management shares or any other territorial project (Daydie, 2016, pp.354–356).[37] Added to the financial resources derived from the taxation of production facilities and associated with the number of jobs created when setting up a RE production project, participation in the capital of an electricity SPC can constitute a virtuous budgetary circle for the communities involved.

The minority but mandatory presence of private shareholders is another attraction of the formula. Thus, if the private shareholders cannot control the company, they finance the project to a certain extent and bring indispensable technical expertise, know-how and entrepreneurial logic. As essential partners for RE production projects, the importance of their presence in the capital of electricity SPCs should, however, be put into perspective. Thus, out of the 72 entities studied, their shareholding is on average only 31%. Like other companies in the local public sector,[38] electricity production SPCs are therefore, in reality, more a means of local governance than a real place of partnership with the private sector. A more precise analysis of the capital partnerships also makes it possible to measure the artificiality of the mixed nature of the capital of these SPCs. Thus, more than 50% of these LPEs have a local public shareholding of between 70% and 85%. However, 34% of them have between 50% and 60% of their capital held by local authorities, which helps to mitigate some of the criticism of an excessive imbalance in the public–private partnership relations present in these SPCs. As for the number of private shareholders in these entities, it appears that there are practically as many SPCs with a single investor (21) as there are with two (18), three or even more (21). It seems therefore that, by all accounts, there is no privileged type of capital allocation in these LPEs (Figure 6.1).

The identity of public shareholder partners is traditional investors, which come from the public sector: the *Caisse des dépôts et consignation,* national and corporate banks, regional investment funds, EDF, LDCs and other SPCs. Thus 24 electricity SPCs (including two LDCs) out of 72 have shareholders, half of which are considered private, other SPCs generating, supplying or marketing electricity and the other half SPC-LDCs. This shareholding represents an average of

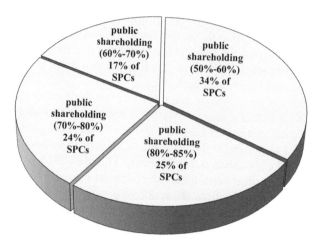

Figure 6.1 Breakdown of majority public shareholding in SPCs.

22.9% of the capital of these entities and varies between 0.4%[39] and 49.9%.[40] These cross-shareholdings are therefore not anecdotal, and they clearly raise the question of the mix of capital in companies that "presuppose a real private partnership" (Ministère de l'intérieur et de l'aménagement du territoire, 2007, p.58). In the final analysis, and in view of the control exercised by local authorities over electricity SPCs, the reasons for the presence of the private sector in the capital are questionable, all the more so as minority shareholders,[41] these operators can only draw few dividends from companies that, moreover, have the reputation of distributing only a few.[42] It appears that, in fact, these shareholders are only complementary fund providers, "sleeping partners" (Brameret, 2011, §284)[43] whose final goal is to obtain "a ticket to enter the public sector [in order to] develop [their] activity through a privileged information channel" (Fontowicz, 2006, p.98) (Figure 6.2).

Local public companies have little presence in the electricity sector, even though in theory they are an adequate legal instrument.

The local public company, an appropriate but under-used shareholder partnership solution

As a public limited company whose capital can only be held by local authorities and their groupings, the LPC allows two or more "authorities to cooperate on a specific issue or a specific function without these authorities losing that power" (Carles, 2013, p.1099). This tool is therefore a good way for local authorities to implement energy policies together, operate and develop RE facilities, delegate project management, develop consulting and study functions (Fournon, 2010). The public shareholding, which can only be total, also enables local authorities

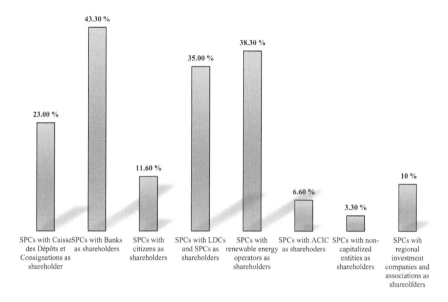

Figure 6.2 Involvement of the various types of private shareholders in the capital of 60 electricity SPCs as a percentage of the number of SPCs.

to exercise control over the structure that must be dedicated to it that is equivalent a priori to that which they exercise over their own departments. Under these conditions, the contractual relations established between the local authorities and the LPC on the occasion of a public service delegation or a public contract are excluded from the scope of the application of the rules on competitive bidding.[44] In view of these advantages, which also explain the very significant increase in the overall number of LPCs since 2010,[45] the presence of only five LPSs in the electricity sector raises questions. In reality, two explanations can be put forward to explain this situation. First, the absence of private shareholders in the capital of an LPC means that the financing of the entity is based on the profitability of its activities and on the resources of the local authorities. LPSs are therefore dependent on public funds, the limited nature of which is well known. Moreover, since they can only act on behalf of their shareholders, only on their own territory and without the possibility of taking stakes in other companies, LPSs are unable to diversify or become subsidiaries. They are therefore much less versatile than SPCs. Second, the lack of any real cooperation between the authorities that are shareholders in the LPC may have legal consequences that encourage them not to do so. A study of the capital of these entities (all activities combined) shows that 92% of them are dominated by a single shareholder holding more than half of the capital (Brameret, 2014, pp.159–166). Electricity LPCs are in the same situation: one is dominated by an ultra-majority shareholder and three others have a single shareholder owning a little more than 50% of the capital.[46] The problem here, however, lies in the distribution of the minority capital. While two electricity LPCs have a more or less balanced distribution of capital,[47] two others have a fragmented shareholding among a multitude of local public actors.[48] When viewed in the context of the requirement of "real analogous control",[49] this fragmentation raises doubts about the reality of its existence.[50] Thus, for example, it seems unlikely that the local authorities that are shareholders in LPC *Horizon* and *Bois énergies renouvelables*, which hold between 0.06% and 17% of the capital, will in practice exercise over these two LPEs a supervisory and management power equivalent to that which they exercise over their own departments, unless, of course, statutory and extra-statutory instruments have arranged for this control.[51]

In short, although the LPC might at first seem, on account of its legal character-istics and its purpose, to be "perfectly suited to the growing involvement of local authorities in the development of renewable energies" (Mauger, 2017, p.522),[52] it has to be said that its use in the local electricity sector does not, in its current state, meet the needs of local authorities.

While majority local public shareholding is therefore the preferred means of intervention of local authorities in the field of electricity distribution and produc-tion, the possibility now offered to them to hold a minority stake in various struc-tures working in the RE sector opens up new opportunities for them.

Minority public shareholding and local electricity policy, new perspectives

Since the law of 1 July 2014 created the SPSPCs (Devès, C. 2019) and the law of 17 August 2015 on energy transition, local authorities have the right to become, like the state, minority shareholders in public limited companies. This trivializa-tion, which can also be interpreted as a sign of the deepening of free administra-tion, is however currently only slightly developed in the electricity sector. Its novelty, but above all the unsuitability of the semi-public company with a single purpose (SPSPC), which is its vehicle, to the energy segment, explains this situa-tion in the first instance.

However, the acquisition of minority shareholdings in the capital of private renewable energy production companies and the creation of hydropower semi-public companies (HSPC) constitute other local tools, promising involvement in energy activities.

Single-purpose semi-public companies, a company with legal specificities, ill-suited to the local electricity sector

Composed of at least two shareholders, including a local authority or a group of local authorities and an economic operator (Art. L 1541-1 LGC), the SPSPC, which is a public limited company, has the specificity of being created following a call for tenders from economic operators likely to be interested both in enter-ing its capital and in the execution of the contract which simultaneously sets its corporate purpose and its field of activity.[53] The call for competition is made only at the stage of the choice of economic operators wishing to become shareholders of the SPSPC and therefore prior to its creation. The contracts binding the local company and its shareholders then escape any subsequent call for competition. The SPSPC is also the first local public company in which local authorities can be minority shareholders. However, their shareholding may not fall below 34%, that is, below the blocking minority, and may not exceed 85%.[54] While this new formula of shareholder partnership can strengthen the position of local authorities in the field of electricity production, the SPSPC is also a means of compensating for the scarcity of local public resources. The private shareholder, when it is a majority shareholder, can then ensure the financing of the creation and management of electricity production facilities. The renewable energy sector is indeed booming and a source of income because there remains an obligation for Enedis and the ELDs to buy out non-eligible customers at the price set by the state within the framework of the regulated sector. Private investors and public authorities can thus see SPSPCs as interesting tools. However, in practice, it appears that no SPSPC has been created to date in the field of electricity generation. The complexity of an electricity production project, which often involves a large number of players at various stages, is a first obstacle to the use of this LPE, particularly because of the unique, immutable and limited nature of its corporate purpose.[55] It is therefore not certain that "such a semi-public company (can), for example, build works and then operate them, [because] in this case is there still uniqueness in the operation?" (Devès, 2014, §22). If it is not absolutely impossible to set up such a company in the electricity sector, it is then possible to wonder about the interest of using a tool that is still poorly used and whose contentious prospects are moreover relatively unknown. Minority local shareholding can also pose difficulties if local authorities are not able to effectively control their private partners. If this remark is addressed to the classic SPCs,[56] it may also be addressed to the SPSPC. The asymmetry of the situation in which local authorities find themselves in certain public service delegations or public-private partnerships risks being amplified in the context of such a PLC, especially if certain big water or electricity distribution operators invest in the segment. This question of the balance of power could also take on a particular aspect when a SPSPC malfunctions. Would local authorities really dare to impose penalties on their private partners in the event of poor execution of the mission covered by the initial contract of their electricity SPSPCs? (Vanneaux, 2016, pp.42–49).

While the electricity production sector has thus far been marked by the absence of SPSPCs, local and regional authorities seem to be taking stakes in private renewable energy operators.

The path of public minority shareholdings in private renewable energy companies

Traditionally, as a rule, local authorities could not acquire shares in the capital of commercial companies and any other profit-making organization not intended to operate municipal services or activities of general interest.[57] Article 109 of the law of 17 August 2015 overturned this historic principle and authorizes local authorities, on the basis of a local council's decision, to participate in the capital of public limited companies or simplified joint-stock companies whose corporate purpose is the production of renewable energy through facilities located on their territory, or on territories located nearby and participating in the energy supply of their territory.[58] The legal framework of this new minority shareholding seems at first sight clear and not very restrictive, at least for the municipalities, their groupings and the *départements*. Indeed, the only limitation relates to the corporate purpose of the company concerned, which must be linked to renewable energy production. Moreover, public authorities are not required to meet any capital ownership threshold and only minority shareholders in the *région* are subject to limits.[59] The legal regime established by the aforementioned Article 109 even seems to provide local authorities with more autonomy by allowing them, through their minority shareholders, to free themselves from the geographical scope of their territorial competences even though it is strictly imposed in the context of the majority principle.[60] Indeed, the legislator specifies that shareholdings may be acquired in companies located in areas close to the local authorities that are shareholders. However, there are some obstacles in practice. First, the acquisition of a share in a renewable energy production company outside the territory of the investing local authority is only possible if it is justified by a need for energy supply. However, if the concept of local territory is not defined by the texts, the inability to prove that such an offshore installation produces the energy necessary for the public entity risks preventing the capital transaction. Once the electricity is "injected into the network [remaining] physically untraceable" (Noël and Orier, 2015, p.41), it is impossible to show that the generating company located outside the territory of the minority shareholder local authority, participates in supplying it, given that the latter term, "supply", is also vague. Second, only companies whose corporate purpose is the production of renewable energy may acquire a minority shareholding. According to Article 109 of the law of 17 August 2015, companies specializing in renewable energy development and/or energy management projects are therefore excluded a priori from all shareholdings. This represents a significant limit to the scope of activities of local and regional authorities in areas particularly relevant for energy transition. In addition, the question of their ability to acquire capital in companies whose corporate purpose includes both renewable energy production and one of the two activities previously mentioned is therefore raised. A final remark, as a warning, can finally be made with regard to these new opportunities for minority public shareholding. While this does not automatically qualify as state aid under EU and national law,[61] local authorities that take part in the capital of renewable energy production companies under these conditions must behave like investors in a market economy.[62] Thus, "beyond the ecological interest of the project, [the local authorities] must ensure that [their] financial interest is not compromised by the principle of this equity investment" (Noël and Orier, 2015).[63] In other words, these local minority shareholdings must be profitable. From this point of view, local minority shareholdings in the electricity sector are no different from those found in other types of LPEs. Local authorities attracted by this new economic intervention modality, in a sector that is itself still experimental, should therefore be cautious. This is even more true after the recent introduction of two new partnerships based on a minority entry into the capital initiated this time by other actors than local authorities.

Article 111 of the law of 17 August 2015[64] thus introduced a mirror mechanism complementing that set up by Article 109. It now allows joint-stock companies, SPCs or cooperatives companies[65] with a RE production project, when they are created or their capital changes, to sell part of their capital, notably to local authorities and their groupings in the area where the project is located or close to this location. Local authorities are thus once again offered the opportunity to invest in green electricity production while limiting their financial involvement, as their shareholding is necessarily a minority one. This is particularly the case when opening the capital of cooperative societies of collective interest (CSCI), knowing that each shareholder has only one vote regardless of the amount of capital held.[66] The CSCI can also be a relevant special-purpose tool for local public shareholders. Indeed, as for any cooperative company, it enables local authorities to be associated with the citizens who are themselves shareholders, which is important in terms of membership of a RE project. Moreover, the purpose of the CSCI, which is limited by the legislator to the production or supply of goods and services of collective interest that have a social utility,[67] is particularly well suited to a RE production project. The only possible obstacle to the inclusion of local authorities in the capital of CSCI is in the case it has a profit-making objective.[68] But in any case, and beyond the CSCI solution, the equity investments proposed by Article 111 of the TE law are subject to two conditions. First, only local authorities located on the territory or near the location of the installation can respond to calls for capital, and second, only the acquisition of shares in a RE production project company is possible. However, as already mentioned, the notion of "local area" and that of "project company" are not defined verbatim. These terminological imprecisions may not, however, hold back public authorities, which, by playing on these ambiguities, may then attempt to develop their minority public shareholding both materially and geographically. Ultimately, by taking minority stakes in companies outside their territory, local authorities are once again broadening their territorial scope thanks to the development of renewable energies.

The context in which the last hypothesis of minority shareholding by local and regional authorities takes place is very different from the previous ones, in particular, because it complements that of the state in the context of a SPC with a new status.

The hydropower semi-public company, a singular public participation company

Qualified as "Houille Blanche" ("white coal"), the electricity produced by rivers and waterfalls has for a very long time been mainly exploited through concessions[69] granted to EDF, the historic operator, and automatically renewed by the state. This concession system, which is not in fact subject to the traditional obligations of competitive tendering when they are reallocated, despite the fact that it is a delegation of public service,[70] and despite the law of 29 January 1993 relating to the prevention of corruption and the transparency of economic life and public procedures,[71] has been severely criticized by the European Commission.[72] On 22 November 2015, the Commission thus served notice on France to bring its legislation into conformity with Article 106 of the Treaty on the Functioning of the European Union (TFEU).[73] The law of 17 August 2015 and the various texts adopted in 2016 on concessions[74] provide an opportunity to correct the situation and allow local authorities to become minority shareholders in the hydropower semi-public companies (HSPC) created on this occasion. Thus, when renewing such a concession, the state can now grant it to a public limited company set up on an *ad hoc*[75] basis comprising at least one economic operator and local authorities bordering the watercourses. This HSPC is a single-operation company set up for a limited period with a view to the conclusion and execution of the concession, the purpose of which is the development and operation of one or more installations constituting a chain of hydraulically linked facilities. Its corporate purpose may therefore not be modified throughout the duration of the contract. As public shareholders must hold between 34% and 66% of the company's capital, they are therefore guaranteed a blocking minority. The economic operator must also hold at least 34 % of the voting rights or capital of the HSPC. The selection of the economic operator is made at the same time as the concession contract, which the HSPC will operate, is put out to tender. The legal regime of the HSPC, therefore, appears to be modeled on that of the SPSPC created a few months earlier. However, a number of particularities distinguish them from it. For example, not only is the entry into the capital of local authorities a prerequisite for the selection of the private shareholder(s), but it is also subject to authorization by the state. The terms and conditions of the association between all the public partners within the company are also the subject of an agreement prior to its creation.[76] This agreement covers, in particular, the share of capital that the public partners wish to hold and the methods of governance and control of the company's activity. These latter characteristics make the HSPC a *sui generis* (Mauger, 2017, p.529)[77] SPSPC that does not express any desire for decentralization and can hardly be described as a local public company. In reality, the HSPC shows "very clearly [the State's will] to maintain public control in this sector at a time when the opening to competition is supposed to bring new concessionaires" (Boiteau, 2017, p.62).[78]. However, as things stand at present,[79] the HSPC is certainly not the seat of a balanced cooperation between the various types of shareholders, especially as the nature of the economic operator partner is not defined by Article 118 of the law of 17 August 2015, and there is great fear that EDF, today's incumbent, will remain the partner.

Conclusion

"Accelerating the energy transition" (Fedepl, 2019): here is a summary of the role now devolved to local public companies and local authorities in the development of renewable energies and the fight against climate change. The local public sector is undeniably a key player in this field and its physiognomy ultimately reflects the way in which local authorities assume these new environmental functions. While LPEs have always worked in one way or another in the narrow electricity sector, the various legislations that have accentuated the role of local public actors, particularly in RE production, have had undeniable consequences on these operators. The panorama of electricity distribution and production LPEs, the photograph of their shareholding and the overview of their activities demonstrate this perfectly. Thus, not only has the number of LPEs continued to grow over time, but paradoxically only the SPCs seem to have won the favor of local authorities, whereas, since 2010, they have also been able to use LPCs and SPSPCs. While experience in the use of this tool, its flexibility and the control of capital that it induces explain this situation, the new opportunities to become a minority shareholder of private electricity production companies resulting from REs are however likely to reconfigure the local public shareholding and thus the sector.

A quick look at the activities of the electricity EPLs reveals two things. Firstly, the interweaving of local authorities and their groupings' functions resulting from the texts[80] has the consequence of being a factor of diversification of the public shareholding of these companies. Thus, for example, the production of green electricity is a function of the municipalities considered here as the basic territorial level (they are also responsible for the distribution and contribution to the energy transition and the control of energy demand[81]), which they can transfer to intermunicipal unions, unions of *départements* and public establishments for intermunicipal cooperation (PEIC).[82] In addition, like the *départements* and *régions*, they can now produce electricity from renewable energy sources on their own initiative.[83] Some SPCs will then have a local authority or a grouping as majority shareholder and others will have an association of several public entities. As Figure 6.3 shows, the public shareholding of these LPEs can be quite heterogeneous in this respect.

It should be noted, however, that there is a very strong presence of municipalities and intercommunal and departmental unions in the capital of the SEMLs of electricity production and energy management. The municipalities in 27% of the cases and the trade unions and PEIC in 20–22% of the situations are the leading shareholders of these local public companies, knowing that 13 intermunicipal and unions of *département* syndicates, four PEIC and three municipalities alone concentrate between 70 and 85% of the capital of these companies. These actors are therefore undoubtedly the territorial levels of principle and operation.

Second, and finally, although the corporate purpose of the LPEs and therefore their field of intervention is generally in line with the legal jurisdiction of the public shareholders, it is not uncommon for these companies to combine activities. Thus, eight out of 18 LDCs carry out three activities, some of which have little to do with their core function, energy (road signs or public lighting);[84] 31 SPCs in renewable energy production and control of energy consumption out of 52 have more than three activities, 12 have two and 9 have only one. These ancillary

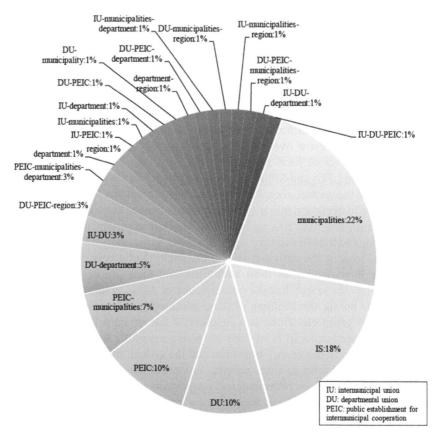

Figure 6.3 Recurrence of combinations of local public shareholders in electricity SPCs -
An analysis of 70 SPCs.

functions are often linked to the operation of production units[85] and the support of other structures working in the sector.[86] On the other hand, the production of wind or photovoltaic electricity, which is supposed to be at the heart of territorial RE projects, is only rarely found in the corporate objects of the SPCs.[87]

In the final analysis, the LPE analysis is a concrete and original means of evaluating the operational management of local electricity policy by local authorities and, more generally, of assessing the way in which they have tackled the energy issue over the last ten years.

The use of these structures, the choice of their form, shareholding and activities, which are real Swiss knives in renewable energy production, clearly shows that more attention is being paid to an issue whose importance "has [even] risen to the top of the agenda in terms of the autonomy of local authorities" (Boutaud, 2013, p.195).

Notes

1 We would like to thank the Fédération of Local Public Enterprises (Fedepl) for giving us access to its documentation and without whom this work could not have been carried out.
2 Art. L 333-1 Energy Code (En. C.).
3 Art. L 211-2 En. C.
4 Law n° 2009-967 of 3 August 2009 on the implementation of the Grenelle Environment Forum, Official Journal of the French Republic (OGFR) of 5 August 2009, p.13031(Grenelle I); Law n°2010-788 of 12 July 2010 on the national commitment to the environment, OGFR n°0160 13 July 2010, p.12905 (Grenelle II).
5 As owners of the distribution networks, these local authorities operate at their own level this activity, which is qualified as a local public service., see for ex. Auby, J.-F. "Les collectivités locales et le nouveau droit de l'énergie", *Dr. Adm.*, n° 12, December 2003, chron. 21.
6 The following will therefore be excluded from the category of local public enterprises, and therefore from the present study, non-capitalized entities (public corporations and public establishments)
7 Art. L 1531-1 Local Government Code (LGC); Art. L 1541-1 LGC.
8 The Finance Act of 16 April 1930 will also make the department a granting authority.
9 Art. 1 law n°46-628 of 8 April 1946, OGFR 9 April 1946, p.2951.
10 NDDs are currently also known as local distribution companies (LDCs). The two terms can be used interchangeably.
11 R. Allemand counts more than 500 (Allemand, R. (2007) "Les distributeurs non nationalisés d'électricité face à l'ouverture à la concurrence", in Annuaire des collectivités locales. *Les collectivités territoriales et l'énergie*. Paris: CNRS, Tome 27, p.31.
12 Article 6 of Act n°. 49-1090 of 2 August 1949 amending articles 8, 23, 46 and 47 of the Act of 8 April 1946 on the nationalization of electricity and gas, OGFR 6 August 1949, p.7712.
13 See Gabillet, P. op. cit., pp.124–131.
14 Thus was created in 1955, *Enercal*; in 1986, *GEG*; in 1989, *Sem énergie*; in 1990, *Humelec*; in 1991, *EDSB*; in 1992, *Semelec 63*; in 1993, *ESC*; and finally in 1997, *EDM*.
15 This increase is explained in particular by the adoption of Act n° 83-597 of 7 July 1983 on local semi-public companies (modernized by Act n°. 2002-1 of 2 January 2002).
16 Law n° 2000-108 of 10 February 2000 on the modernization and development of the public electricity service, OGFR 11 February 2000, p.2143.
17 Law n° 2003-8 of 3 January 2003 on the gas and electricity markets and the public energy service, OGFR 4 January 2003, p. 265; Law n° 2004-803 of 9 August 2004 on the public service of electricity and gas and on electricity and gas companies, OGFR 11 August 2004, p.14256; Law n° 2006-1537 of 7 December 2006 relating to the energy sector, OGFR 8 December 2006, p.18531.
18 Art. L. 111-57 En. C.
19 Belot, C. (4 June 2013) *Rapport d'information n° 623*. Sénat. Délégation aux collectivités territoriales et à la décentralisation. session ordinaire 2012–2013, p.19.
20 Law n° 2005-781 of 13 July 2005 on the program setting out energy policy guidelines., OGFR 14 July 2005, p.11570.
21 See Billet, P. "Premières impressions sur la loi de programmation relative à la mise en œuvre du Grenelle de l'environnement", *JCP A*, 19 October 2009, pp.35–47.
22 See, for example, AJDA, *Dossiers spéciaux Loi Grenelle II*, 1st Part n°30/2010 (20 September 2010), 2nd Part n°32/2010 (4 October 2010).
23 As in the case of municipalities and their groupings, the electricity produced in this way can only be consumed by the municipality itself or sold to EDF or an NND (art. 10 of law n° 2000-108 of 10 February 2000, op. cit.).

24 Law n° 2015-992 of 17 August 2015, OGFR 18 August 2015, p.14263.
25 Law n° 2010-559 of 28 May 2010 for the development of local public companies, OGFR 29 May 2010 p.9697.
26 Law no. 2014-744 of 1 July 2014 allowing the creation of single-transaction semi-public companies, OGFR 2 July 2014, p.10897.
27 Art. 109 à 111 et art. 118, Law of 17 August 2015, op. cit.: Coded to art. L 314-27 Enr. C. and L 521-18 Enr. C.
28 One hundred and seventeen (117) LPEs operate in the energy sector (Fedepl. *EPLScope* 2018, p.6. Available at: https://www.lesepl.fr/zoom/eplscope-2018/.
29 These five SPCs were created between 2013 and 2018.
30 This figure rose from nine in 2000 to 63 in 2019 (data taken from the study of the status of LPEs compiled by the Fedepl).
31 This figure remains low even though article 29 of Law n° 2004-803 of 9 August 2004 (op. cit.) has given the possibility for non-capitalized entities (public corporations) to group together in this form.
32 Art. 24, Law n° 2010-1488 of 7 December 2010 on the new organization of the electricity market (the so-called NOME Act), OGFR 8 December 2010, p.21467.
33 Art. L 1522-1 and -2 LGC.
34 Art. L 1521-1 LGC.
35 Art. L. 1524-5 LGC.
36 Art. L 314-1-13 En. C.
37 Daydie, L. op. cit., pp.354–356.
38 See Antona,F. (2011) *L'économie mixte: un instrument de l'action locale*, Doctoral Thesis, Université Lille 2, p.448 and seq.; Brameret, S. op. cit., §265 and seq.
39 *Oregies* and *Sylneva collectivités*.
40 *EDC- SAEML Énergie développement Cervières* (Gabillet, P. op. cit., p.194).
41 Out of the 63 private equity companies analyzed, only 15 have a private shareholding of between 40% and 50% of the capital.
42 Peltier, M. (2007), *La participation des collectivités territoriales au capital des sociétés*. Aix-Marseille: PUAM, n°183 and seq.
43 Brameret, S. op. cit., §284.
44 Direction générale des collectivités locales, Circ. 29 avr. 2011, n° COT/B/ 11/08052/C, relative au régime juridique des sociétés publiques locales et des sociétés publiques locales d'aménagement; Commission of the European Communities (CEC), 17 July 2008, aff. C-371/05, Commission of the European Communities v/ Rép. Italienne (Cne Mantoue); CEC, 18 January 2007, aff. C-220-05, Jean Auroux e.a. v/ Cne Roanne; CEC, 19 April 2007, aff. C-295/05, Asociación Nacional de Empresas Forestales (Asemfo).
45 The number of LPSs increased from 85 in 2012 to 389 in 2019 (v.Fedepl, EPLscope, www.lesepl.fr).
46 The Réunion holds 81.78% of the capital of LPC *Horizon*; 51% of the capital of LPC *SEER Grigny-Viry-Paris* belongs to SIPPEREC (intercommunal syndicate); the mixed union of Martinique holds 52% of the capital of the LPC *Martinique énergies nouvelles*; 50.33% of LPC *Bois énergie renouvelable* shares are owned by the city of Lorient. On the other hand, the composition of the capital of LPC Chartres *Métropole énergies* remained unknown.
47 In the *SEER*, one municipality holds 15% of the capital, the other 34%. In the LPC *Martinique énergies nouvelles*, two agglomeration communities each own 12% of the capital, an intercommunal syndicate owns 18% of the shares and another syndicate 6%.
48 In *SPL Horizon*, six shareholders hold between 1% and 4.99% of the capital (15 municipalities, one public interest group and one mixed union hold between 0.06% and 0.67% of the shares). In LPC *Bois énergies renouvelable*, 12 municipalities hold 0.33% of the capital, one other 0.6%, three municipalities 9.3% and Lorient agglomeration 17% of the capital.

49 The advantage of creating an LPC for local authorities lies in the fact that the contracts they enter into with this type of company are not bound by public procurement law and are therefore not subject to competition law. However, in order to benefit from this hypothesis, the case law of the Court of Justice of the European Union and French law requires that the economic operator has an activity that is more than 80% dedicated to its shareholder communities and that the latter exercise control over it similar to their own departments. This control must be real, effective and demonstrable. In this context, a minority holding of capital, and a fortiori an ultra-minority holding, does not a priori allow such control to be exercised and therefore the company in relation to two is not considered to be *in-house*.

50 See in this sense CE, 6 November 2013, n° 365079, Commune de Marsannay-la-Côte.

51 Fedepel (2013) *SPL et contrôle analogue,* Paris: Fedepl, coll. Mode d'emploi.

52 Mauger, R. (2017), *Le droit de la transition énergétique, une tentative d'identification,* Doctoral Thesis, Université Montpellier, p.522.

53 Art. L 1541-2 LGC.

54 Art. L 1541-1 III al 3 LGC.

55 The SPSPC ceases to exist at the end of the missions provided for in the contract, i.e., at the end of its execution.

56 See in this sense: Cour des comptes, *Les sociétés d'économie mixte locales, un outil des collectivités à sécuriser,* May 2019, p.86, p.15. Available at: https://www.ccomptes.fr/system/files/2019-05/20190527-societes-d-economie-mixte-locales.pdf.

57 Art. L 2253-1 LGC.

58 Art. L 2253-1 al 2 LGC for municipalities and their associations; art. L 3231-6 LGC for departments; art. L 4211-1 LGC for regions.

59 It can thus be exercised only within the framework of its economic development jurisdiction (art. L. 4211-1 8° bis LGC), to be established only in RE companies carrying out all or part of their activities on the regional territory and not to have the effect of increasing the share of capital held, directly or indirectly, by public entities to more than 50% or to increase its own share to more than 33% of the company's capital and for a maximum of 33% of the capital of the latter (art. R. 4211-1 to -5 LGC).

60 Cour des comptes, op. cit., p.30; for a strict application of the territoriality principle to LPSs and SPSPCs see art. L. 1531-1al 3 LGC et art. L. 1541-1 LGC.

61 See in this sens: CEC, 14 November 1984, SA Intermill c/ Commission, aff. 323/82.

62 Available at: https://blog.acpformation.fr/2017/02/28/modalites-dinterventions-directes-collectivites-territoriales-capital-dentreprises-privees-a-suite-innovations-de-loi/

63 Noël, M. and Orier, J. op. cit.

64 Art. L. 314-28 En. C.

65 Law n° 47-1775 of 10 September 1947 on the status of cooperation, OGFR 11 September 1947, p.9088.

66 Art. 19 septies et octies, of the Law n°47-1775, op. cit.

67 Art. 19 quinquies and seq. of the Law n° 47-1775, op. cit.

68 Dondero, B. and Le Cannu, P. (2019) *Droit des sociétés.* 8ème éd. Paris: LGDJ, Précis Domat, §18.

69 See art. 1er of the law of 16 October 1919 on the use of hydraulic energy.

70 CE, sect. des travaux publics, avis 28 September 1995, n°357262 et 357263.

71 Law n° 93-122 of 29 January 1993 relating to the prevention of corruption and the transparency of economic life and public procedures, chap. IV and seq., OGFR 30 January 1993 p.1588.

72 See, for example, European Commission press release of the 21 December 2004 (IP/04/1520) and 13 July 2005 (IP/05/920).

73 According to this article "In the case of public undertakings and undertakings to which Member States grant special or exclusive rights, Member States shall neither enact nor maintain in force any measure contrary to the rules contained in the Treaties, in par-

ticular to those rules provided for in Article 18 and Articles 101 to 109. Undertakings entrusted with the operation of services of general economic interest or having the character of a revenue-producing monopoly shall be subject to the rules contained in the Treaties, in particular to the rules on competition, in so far as the application of such rules does not obstruct the performance, in law or in fact, of the particular tasks assigned to them. The development of trade must not be affected to such an extent as would be contrary to the interests of the Union".

74 See order n°2016-65 of 29 January 2016 on concession contracts (OGFR 30 January 2016); Enforcement Decree n°2016-86 of 1 February 2016 relating to concession contracts (OGFR 2 February 2016); Decree n°2016-530 of 27 April 2016 relating to hydropower concessions and approving the model specifications applicable to these concessions (OGFR 30 April 2016).

75 Art. L 521-18 En. C.

76 Art. L 521-19 En. C.

77 Mauger, R. op. cit, p.529.

78 Boiteau, C. op. cit., p.62.

79 None of these companies have yet been created. However, on 8 February 2019, the Savoie Departmental Assembly voted in favor of the creation of a departmental hydro-power HSPC, which would intervene on Savoie concessions should the state decide to open them up to competition in the future.

80 ORCE (November 2016) *Réforme territoriale: l'exercice des compétences énergie dans les territoires*. ENJ06, Table 3, p.22 and Table 4, p.23; AMORCE (February 2017) *Les compétences "Énergie-climat" des departments*. ENJ07, Annex et Table, p.14; Daydie, L. op. cit. Annex 1 et Tables, p.511 and seq.

81 Art. L 1111-2 LGC.

82 Art. L 2224-33 LGC.

83 Art 88 of the law n°2010-788 of 12 July 2010, op. cit.

84 Sablière, P. (2014–2015) *Droit de l'énergie*. Paris: Dalloz, coll. Dalloz Action, §815.51-.56; ANROC (October 2012) *Les entreprises locales de distribution d'électricité, l'originalité d'un service local de proximité-Panorama, enjeux, perspectives*, p.13.

85 Acquisition, financing, construction, development and operation of renewable energy production facilities.

86 Accompaniment and development of RE projects and equity investments, sharehold-ing and investment in companies specialized in the production of RE and RE project companies.

87 These conclusions are based on an analysis of the only corporate objects that could be found.

References

Belot, C. and Juilhard, J.-M. (2006, June 28) *Rapport d'information n°436*, Sénat. Délégation à l'aménagement et au développement durable du territoire sur les énergies locales, session ordinaire 2005–2006, p. 39.

Bizet, J.-F. (2012) *Entreprises publiques locales, SEM, SPLA, SPL*. 2nd ed. Paris: Lamy, Coll. Axe Droit, p. 15.

Boiteau, C. (2017) *Les entreprises liées aux personnes publiques*, RFDA, 2017, n°1, pp. 57–67.

Boutaud, B. (2013) 'Les énergies renouvelables, énergies des collectivités territoriales?' in Annuaire des collectivités locales (ed.) *Collectivités territoriales et énergie: ambitions et contradictions*. Paris: Le Moniteur, Tome 33, p. 195.

Brameret, S. (2011) *Les relations des collectivités territoriales avec les sociétés d'économie mixte locales. Recherche sur l'institutionnalisation d'un partenariat public-privé.* Paris: LGDJ, coll. Bibl. dr. pub., Tome 271, §219 and seq.

Brameret, S. (2014) 'L'actionnaire type d'une société publique locale', *Rev. Adm.*, March-April 2014, n°398, pp. 159–166.

Carles, J. (2013) 'La société publique locale, un outil adapté à l'évolution des missions du service public local', *RFDA*, 2013, p. 1099.

Da Rold, J. (2008) *Les sociétés d'économie mixte locales: acteurs et témoins de politiques urbaines et territoriales*, Doctoral Thesis. Université Bordeaux III, p. 181.

Daydié, L. (2016) *Personnes publiques locales et énergie*, Doctoral Thesis. Université de Pau et des pays de l'Adour, p. 34 and seq., p. 239 and seq.

Devès, C. (2014) 'La loi n° 2014-744 du 1er juillet 2014 permettant la création de sociétés d'économie mixte à opération unique: innovation ou fuite en avant?', *JCP A*, pp. 38–39, §22.

Devès, C. (2019) 'Société d'économie mixte à opération unique', *JCl. Collectivités territoriales*, fasc n°481, 4 décembre 2019.

Fedepl (2019) *Accélérer la transition énergétique avec les EPL.* Paris: Fedepl, coll. Stratégie.

Fontowicz, L. (2006) 'La performance de gestion des SEM' in Guerard, S. (ed.) *Regards croisés sur l'économie mixte: approche pluridisciplinaire droit public et droit privé.* Paris: L'Harmattan, p. 98.

Fournon, A. (2010) 'Le recours aux sociétés publiques locales (SPL) comme nouvel instrument du développement des énergies renouvelables pour les collectivités', *Gaz. Pal.*, September 25, p. 34.

Gabillet, P. (2015) *Les entreprises locales de distribution à Grenoble et Metz: des outils de gouvernement énergétique urbain partiellement appropriés*, Doctoral Thesis. Université Paris-Est, p. 124.

Jourdan, F. (2012) 'Une société d'économie mixte pour favoriser la transition énergétique en île-de France', *JCP A*, march 26, n° 12, act 197, p. 3.

Marcou, G. (2007) 'Le cadre juridique communautaire et national et l'ouverture à la concurrence: contraintes et opportunités pour les collectivités territoriales' in Annuaire des collectivités locales (ed.) *Les collectivités territoriales et l'énergie.* Paris: CNRS, Tome 27, p. 20.

Mauger, R. (2017) *Le droit de la transition énergétique, une tentative d'identification*, Doctoral Thesis. Université Montpellier, p. 522.

Ministère de l'intérieur et de l'aménagement du territoire, D.G.C.L. (2007) *Le guide des Sociétés d'économie mixte locale.* Paris: La Documentation française, p. 58.

Nicinski, S. (2019) *Droit public des affaires.* 7th éd. Paris: LGDJ, Précis Domat, §774 and seq.

Noël, M. and Orier, J. (2015) 'Participation des collectivités territoriales au capital d'une société privée ayant pour objet la production d'énergie renouvelable', *Contrats Publics*, n°160, pp. 40–43, p. 41.

Vanneaux, M.-A. (2016) 'Les entreprises publiques locales, acteurs des écoquartiers', *Lexisnexis JCl. Actes pratiques & ingénierie immobilière*, jan.-fev.-march, pp. 42–49.

7 Analysis of the value added to local economies by municipal power suppliers in Japan

Kenji Inagaki and Takuo Nakayama

After Japan experienced the 2011 Fukushima nuclear accident, energy policies, which up to that point had been treated as though they fell under the exclusive jurisdiction of the central government, became a major issue for local governments. Many local governments developed their energy visions and policies and took the initiative to develop local renewable energy, though initially, their focus was only on developing renewable sources.

Since the full liberalization of the power retail market in April 2016,[1] Japanese local governments started to enter new power retail businesses by directly or indirectly establishing new municipal power suppliers (MPSs). This means local governments, which have so far focused on introducing renewable energy facilities (e.g., PV on their office building roofs), are taking the next step of leveraging renewable energy through starting their own retailers with the aim of generating benefits in their local communities.

These newly founded MPSs are expected to function as a mechanism to secure local renewable development by procuring power (and heat) from local renewable energy projects and delivering it to customers on their premises. There is also anticipation that MPSs can financially contribute to communities by reducing overall energy expenses and helping local economies. Behind this, there is a reflection that many of the local governments have built large-scale solar plants in their communities, but only to receive property taxes and rental incomes in earnings since the plants were funded and constructed by non-local businesses. For MPSs to not repeat the same footsteps, MPSs should be created in a way to optimize local financial benefits, by, for example, adopting a business structure that facilitates local investment, encouraging local employment and performing operations in-house.

Expectations on the community energy sector

Japanese MPSs' development is largely inspired by the international movement of the civic energy sector. The community energy sector broadly consists of bottom-up energy schemes and municipal business models, with the differentiation made on their origins and stakeholders. Preceding research has mainly studied community-led energy projects by focusing on internal and contextual dynamics of organizations to identify enablers and barriers of renewable development

DOI: 10.4324/9781003025962-11

(Bulkeley, et al., 2014). The discourse to connect the municipal role for energy transitions in a decentralized context is, however, relatively new (Hall et al. 2016,; Nakayama et al., 2016).

Electricity market liberalization has been customarily discussed within the context of creating a level-playing field where existing market players and new entrants compete, with the expected effect of fair and transparent price settings. As part of liberalization, it is often the case to dis-bundle three main parts of electricity utility: power supply, transmission and wholesale. The liberalization process, which in principle requires equal footage, may demand separation of these functions. The wholesale liberalization, *inter alia*, is of particular significance in terms of connecting market decisions and the values represented by a wider "audience": the civil societies and consumers. In this context, a community energy sector concept emerged to encompass citizen, community, cooperative and municipal ownership of energy systems (Hall et al., 2016). The civil ownership structure and institutions, particularly banking and finance, have a strong implication in shaping community energy participation with interaction with regulatory frameworks, for example, feed-in tariffs to regulate or encourage ownership forms (Suwa, 2020).

Municipal business models, though they have their roots in historical energy structure, including local energy initiatives, are rapidly growing. There are about 2800 energy cooperatives in Europe, 1000 of which energy cooperatives are in Germany, but other European countries are following .

Stadtwerke in Germany (and in some other countries in its vicinity) systematically manage and operate the local power supply, gas and heating, transportation, water and sewage, and other social infrastructure projects.[2] The Stadtwerke is understood by Japanese academics and municipal stakeholders as a mechanism to create jobs in local communities, and many of them put a strong emphasis on generating economic circulation in the communities by placing orders with and procuring the renewable-relevant materials from local businesses by offsetting between their revenue from energy operations (e.g., power retailing and district heating) and their non-profitable operations (e.g., transportation and public swimming pool facilities), thereby contributing to maintaining regional infrastructure and services.

These features of Stadtwerke have been predominantly highlighted in Japan and have shaped current MPSs development so that MPSs also have a similar local impact as the German models. Morotomi (2018) argues that the idea of expanding sources of operating revenues other than taxes will become increasingly relevant for Japan's local governments in the future as the country examines how to cover the costs of maintenance and upgrades of aging social infrastructure as well as social security costs associated with an aging society. Meanwhile, there are no prospects for significant increases in revenue from municipal and property taxes due to the country's diminishing population and land value. He proposes the establishment of the Stadtwerke in Japan as in Germany and the development of a business model in which the revenues earned in energy operations are reinvested into other public projects.

The Japanese municipalities, however, often face challenges to convince local citizens, who may not be fully sympathetic to the idea that municipal governments invest tax revenues into their energy businesses. It has often been the case that poor and misleading information is provided to the public, to the extent there is a misconception that renewable energy is only to benefit those who develop and harness profit, especially in cases where the traditional local landscape is ravaged by renewable development.

Measuring "community" economic benefits would be one effective way to convince the communities and let them be interested in local renewable projects. While serving the obvious climate and energy security goals, renewable energy can also be a significant instrument to bring positive regional economic outcomes. The identification of local economic benefits would be a persuasive tool to attract more attention to renewable development from investors and other stakeholders (Suwa, 2020).

This chapter will therefore follow the methodology broadly proposed by the supply chain analysis to recognize regional value added, especially paying attention to the corporate level financial information. First, this chapter organizes 31 new municipal power suppliers in Japan around the following four criteria: investment by local governments, power sources, customers, and supply and demand management. Then, the economic circulation generated by the operations of two new municipal power suppliers is measured by analyzing the value added to local economies. Identifying the local contribution of MPS is yet to be fully materialized, thus the indicative result holds particular importance.

Typology of municipal power suppliers

For MPSs to realize their function to reduce carbon emissions in local communities, and contribute to the local economy, they must establish a business strategy. The business strategy varies, depending on each MPS's principles and stakeholders' expectations. Power supply and demand management, for example, are at the heart of the power retail business and often determine the level of profits. The business strategy also involves an abundance of customer information and provides basic data for developing management strategies. It is important for the MPSs to consider managing power supply and demand in-house, instead of, for example, joining a balancing group with other MPSs.

MPSs must perform operations that require a certain level of expertise not only during the start-up phase but also during the operation phase. Examples of such operations include managing power supply and demand, managing customer relationships, submitting plans to general power utilities and reporting to government authorities. Outsourcing these operations to non-local contractors just because of the expertise required in performing them could reduce community earnings in many cases. New municipal power suppliers need to find ways to broaden the scope of operations to be performed within the community by such means as sharing knowledge with other new power suppliers and employing organizations that provide support and expertise for the establishment of new power retailers.

This chapter, therefore, overlays Japanese MPSs by the following four criteria: (1) investment by local governments, (2) power sources, (3) types of customers and (4) supply and demand management, to understand their current typology and demographic features (Table 7.1).

Table 7.1 List of new municipal power suppliers

Name	Investment, etc.	Renewable energy power sources	Customers	Supply and demand management
Kuji Regional Energy Co., Ltd	Kuji City, Iwate prefecture: 500,000 yen Local companies: 9.5 million yen	Solar power generation facilities in the city	Public facilities, private facilities	N/A
Higashimatsushima Organization for Progress and Economy, Education, Energy	Employees: Residents of Higashimatsushima city, Miyagi prefecture, etc. (An agreement has been signed between the city and the organization.)	Solar power generation facilities in the city	Public facilities Fisheries cooperative, offices, plants	In-house
Kami Power Satoyama Public Corporation	Kami town, Miyagi prefecture: 6 million yen Pacific Power Co., Ltd: 3 million yen	Solar power generation facilities in the town	Public facilities	Outsourced
Yamagata Power Supply Co., Ltd	Yamagata prefecture: 23.4 million yen 18 companies including NTT facilities and local private companies: 46.6 million yen in total	Biomass power facilities (private)Solar power facilities (prefecture owned) Wind power facilities	Public facilities High-voltage private users, low-voltage commercial users	Outsourced
Soma I Grid LLC	Soma city, Fukushima prefecture: 1 million yen IHI corporation: 8.4 million yen Pacific Power: 500,000 yen	Solar power generation facilities in the city	Public facilities High-voltage private users	Outsourced

(*Continued*)

Table 7.1 Continued

Name	Investment, etc.	Renewable energy power sources	Customers	Supply and demand management
Fukaya e-Power Co., Ltd	Fukaya city, Saitama prefecture: 11 million yen Miyama Power HD: 6 million yen Fukaya Chamber of Commerce and Industry: 1 million yen Fukaya city Chamber of Commerce: 1 million yen Saitama Resona Bank: 1 million yen	N/A	Public facilities High-voltage private users	N/A
Chichibu New Power, Co., Ltd	Chichibu city, Saitama prefecture: 18 million yen Miyama Power HD: 1 million yen Local financial institution: 1 million yen	Waste-to-energy power generation facilities (currently under consideration)	Public facilities (power supply scheduled to begin in April 2019)	N/A
Tokorozawa Future Electric Power Corporation	Tokorozawa city: 5.1 million yen JFE Engineering Corporation: 2.9 million yen Hanno Shinkin Bank: 1 million yen Tokorozawa Chamber of Commerce and Industry: 1 million yen	Solar power generation facilities Waste-to-energy power generation facilities	Public facilities Offices, homes (power supply scheduled to begin in or after 2019)	N/A
Nakanojo Electric Power General Foundation (Nakanojo Power Co., Ltd since Nov. 2015)	Nakanojo town, Gunma prefecture: 1.8 million yen V-Power: 1.2 million yen * Nakanojo Power Co., Ltd is 100% owned by Nakanojo Electric Power General Foundation	Large-scale solar power generation facilities (town-owned) Large-scale solar power generation facilities (private: land leased from the town)	Public facilities Homes	Outsourced

Narita Katori Energy Co., Ltd	Narita city, Chiba prefecture: 3.8 million yen Katori city, Chiba prefecture: 3.8 million yen SymEnergy Inc.: 1.9 million yen	Power generation using waste from Narita city's waste disposal facilities, large-scale solar power generation facilities in Katori city (city-owned)	Public facilities	Outsourced
Chiba Mutsuzawa Energy Co., Ltd	Mutsuzawa town, Chiba prefecture: 5 million yen 6 private companies including Pacific Power: 4 million yen in total	Large-scale solar power generation facilities (private)	Public facilities High-voltage private users	Outsourced
Hamamatsu Energy Co., Ltd	Hamamatsu city, Shizuoka prefecture: 5 million yen NTT Facilities: 15 million yen NEC Capital Solutions: 15 million yen 6 local businesses including railways and banks: 25 million yen	Large-scale solar power generation facilities (private) in the city Power generation using waste from waste disposal facilities	Public facilities High-voltage users in the city	Outsourced
Smart Energy Iwata Corporation	Iwata city, Shizuoka prefecture: 5 million yen JFE Engineering Corporation: 94 million yen Iwata Shinkin Bank: 10,000 yen	Wind power generation facilities, solar power generation facilities	Public facilities Private corporate facilities	N/A
Matsusaka Power Co., Ltd	Capital: 8.8 million yen (Matsusaka city, Mie prefecture: 4.5 million yen, Toho Gas: 3.5 million yen, banks and credit unions: 800,000 yen)	Municipal waste-to-energy power generation	Public facilities	Outsourced
Konan Ultra Power Co., Ltd	Konan city, Shiga prefecture: 3.3 million yen 7 private companies including Pacific Power: 5.7 million yen in total	Residential solar power generation systems on leased rooftop	Public facilities High-voltage private users	Outsourced

(*Continued*)

Table 7.1 Continued

Name	Investment, etc.	Renewable energy power sources	Customers	Supply and demand management
Kameoka Furusato Energy Co., Ltd	Kameoka city, Kyoto prefecture: 4 million yen Private companies including Pacific Power and local financial institutions: 4 million yen	Solar power generation facilities in the city	Public facilities	Outsourced
Izumisano PPS General Incorporated Foundation	Izumisano city, Osaka prefecture: 2 million yen Power Sharing: 1 million yen	Large-scale solar power facilities (private)	Public facilities	Outsourced
Ikoma Citizens Power Co., Ltd	Ikoma city, Nara prefecture: 7.65 million yen Osaka Gas: 5.1 million yen 3 other companies: 2.25 million yen in total	Solar power generation, small-scale hydropower generation (city-owned) and collaborative residential solar power generation facilities	Public facilities Private facilities	Outsourced
Tottori Citizen Electricity Co., Ltd	Tottori city, Tottori prefecture: 2 million yen Tottori Gas: 18 million yen	Solar power generation facilities (municipal, private)	Public facilities High-voltage private users, homes	Outsourced
Local Energy Corporation	Yonago city, Tottori prefecture: 9 million yen 5 private companies including Chukai Cable TV System Operator: 81 million yen in total	Power generation using waste from waste treatment facilities Solar power generation facilities in the city (private)	Public facilities	In-house
Nanbu Dan-Dan Energy Co., Ltd	Nanbu town, Tottori prefecture: 4 million yen 4 private companies including Pacific Power: 5.7 million yen in total	Large-scale solar power generation facilities (town-owned) Small-scale hydropower generation facilities (town-owned)	Public facilities High-voltage private users	Outsourced

Okuizumo Electric Power Co., Ltd	Okuizumo town, Shimane prefecture: 20 million yen Pacific Tower: 3 million yen	Small-scale hydropower generation facilities (town-owned)	Public facilities High-voltage private users	Outsourced
Kitakyushu Power Co., Ltd	Kitakyushu city, Fukuoka prefecture: 14.5 million yen 8 private companies including Yaskawa Electric Corporation: 45.5 million yen in total	Waste-to-energy power generation	Public facilities High-voltage private users	Outsourced
Coco Terrace Tagawa Co., Ltd	Tagawa city, Fukuoka prefecture: 2.5 million yen Pacific Power: 2.5 million yen NEC Capital Solutions: 2.5 million yen Financial institutions: 1.2 million yen in total	N/A	Public facilities High-voltage private users	Outsourced
Nature Energy Oguni Co., Ltd	Oguni town, Kumamoto prefecture: 3.4 million yen 7 private companies including Pacific Power: 5.6 million yen in total	Biomass power generation facilities in the town Binary geothermal power generation facilities in the town	Public facilities High-voltage private users	Outsourced
Miyama Smart Energy Co., Ltd	Miyama city, Fukuoka prefecture: 11 million yen Kyushu Smart Community: 8 million yen Chikuho Bank: 1 million yen	Surplus power from residential solar power facilities Large-scale solar power facilities (city-owned) Biomass and other sources procured jointly with other local governments	Public facilities Homes, offices, etc.	In-house

(Continued)

Table 7.1 Continued

Name	Investment, etc.	Renewable energy power sources	Customers	Supply and demand management
Bungo-ono Energy Co., Ltd	Bungo-ono city, Oita prefecture: 11 million yen Denken: 6 million yen Financial institutions, etc.: 3 million yen	Solar power generation facilities (municipally operated) Power shared with Miyama Smart Energy	Public facilities	Collaboration with Miyama Smart Energy
New Energy Oita Co., Ltd	Capital: 20 million yen (Yuhu City in Oita prefecture, Denken, financial institutions and local related businesses (breakdown unknown))	Solar power generation facilities in the city, etc.	Public facilities High-voltage private users, homes	N/A
Ichikikushikino Electric Power Co., Ltd	Ichikikushikino city, Kagoshima prefecture: 5.1 million yen Private businesses, financial institutions, etc.: 4.9 million yen	Power shared with Miyama Smart Energy	Public facilities Homes, etc.	Collaboration with Miyama Smart Energy
Osumi Hanto Smart Energy Co., Ltd	Kimotsuki town, Kagoshima prefecture: 3.35 million yen Kyushu Smart Community: 1.65 million yen	Hydropower, wind power and solar power generation facilities (all privately owned), power sharing with Miyama Smart Energy	Public facilities Homes, etc.	Collaboration with Miyama Smart Energy
Hioki Energy Co., Ltd	Hioki city, Kagoshima prefecture: 1 million yen 14 related local businesses and two individuals: 7.71 million yen Hitachi Power Solutions: 1.49 million yen Kagoshima Renewable Energy Limited Liability Investment Partnership: 20 million yen	Solar power generation facilities in the city Small-scale hydropower facilities in the city (scheduled) Cogeneration facilities (scheduled)	Public facilities Homes, offices, etc.	Outsourced

* Prepared by the authors based on press releases and other official company documents (as of the end of August 2018).

Investment

It is indicative to identify the volume of funds invested into the MPSs to understand the overall size and share of the MPSs in the Japanese electricity market. Also, the degree of investment by municipal governments and local stakeholders can effectively determine the level of dividends and other financial returns to the local economy.

To summarize the MPSs cases, the amount of investment by a local government in a new power supplier generally falls within the range of several million yen, while the investment share varies from 10% to 90%. The total amount of investment is not very large since power retailing, the principal business of new power suppliers, is structured such that power retailers sell electricity obtained from power producers to customers and do not necessarily need to develop and own power generation facilities. Other than local governments, local private businesses tend to be selected to fund many of these new power suppliers to promote the local circulation of capital. In some cases, however, new power companies are funded by non-local businesses. To establish a new municipal power supplier, local governments solicit and select private corporations that will partner with them in a joint investment and technical cooperation in some cases; in other cases, a private corporation submits a proposal for a new municipal power supply project and obtain approval from the local assembly.

Power sources

Not all renewable energy sources and infrastructure contribute to municipal energy and financial autonomy, as often they are owned and operated by the incumbent power utilities. It is, therefore, important to examine to what extent the MPSs supply energy from local sources.

Most of the Japanese MPSs made contracts with local governments to secure electricity, such as from large-scale solar and waste-to-energy power plants that local governments own and operate. Other power sources owned by local governments include hydroelectric plants (public hydroelectric plants) located in dams that were constructed as part of Japan's post-war comprehensive river development projects. As of 1 April 2018, public hydroelectric stations in Japan are owned by 25 prefectures and one city, with a total installed capacity of 2315 MW (Public Electric Utility Enterpriser's Forum). This has limited the availability of public hydroelectric power to the MPSs.

Customers

It is also indicative to observe whether MPSs are supplying energy to public buildings and facilities, or directly to individual households.

Since gaining new customers is the top priority in the power retailing business, being able to secure a certain level of demand associated with local public buildings and facilities (e.g., sewage treatment plants) gives these new power suppliers a significant advantage. In many cases, MPSs are able to secure a contract demand of approximately 5 MW, which is said to be the break-even point from public facilities alone. For this reason, it is common practice for the new power retailers to supply power only to public facilities at first. Some new power suppliers start supplying power to private facilities and residential customers and offering other administrative services after gaining some experience with public facilities and ensuring profitability.

Supply and demand management

For MPSs, whether they outsource or keep supply and demand management in-house creates a significant difference in their overall profitability. "Supply and demand management" is a task consisting of matching power supply with demand and is a key element of the power retail business. Because of the level of expertise required and the time and cost involved, many of the new power suppliers outsource supply and demand management to a private contractor and join its balancing group.[3] (The contractor is a non-local business and has invested in the new power supplier in many cases.) If supply and demand management is outsourced, few employees are needed to run the new power retailer. As a result, a majority of the new power suppliers have only one full-time employee or none.

On the other hand, managing supply and demand in-house offers some advantages, such as accumulating the know-how critical to the power retail business, expanding the opportunities for collaborating with other local governments and (in some cases) increasing profits. For this reason, some of the new municipal power retailers choose to handle supply and demand management in-house. Currently, three new municipal power suppliers manage supply and demand in-house: Miyama Smart Energy Co., Ltd (Miyama city, Fukuoka prefecture), Local Energy Corporation (Yonago City, Tottori prefecture) and Higashimatsushima Organization for Progress and Economy, Education, Energy (Higashimatsushima city, Miyagi prefecture).

Analysis of value added to local economies by new municipal power suppliers

By establishing new power retailers, many local governments seek to generate economic circulation in their communities by producing and consuming energy locally and thereby preventing the outflow of power bill payments from their regions. There is, however, a need to evaluate the impact of MPSs on the local economy.

There are a variety of methodologies developed to identify regional economic impacts of renewable development reflecting the needs for such an evaluation. Jenniches (2018) broadly categorizes four existing methodological approaches, that is, employment ratio, input-output modeling, computable general equilibrium models and supply chain analysis. Following the footsteps of these academic discourses, supply chain analysis and the identification of local context have been applied to renewable energy development in Japan. These include an analysis of technology-specific value added to local communities by renewable power generation projects (Raupach, Nakayama and Morotomi, 2015), and a verification using a case study on a photovoltaic power generation project in Nagano prefecture (Nakayama et al., 2016). These tools for measuring local economies are useful not only for developing energy and climate change policies at the local level but also for enhancing communication among local stakeholders and helping them reach a consensus (ibid.).

The use of financial information to analyze corporate level value added offers practical usefulness, with an obvious reason that the minimum set of data is required, compared with region-wide or technology-wide supply chain analysis. Another more theoretical justification is that, since corporate financial information, in its nature, generically demonstrates financial chains and nexus among different stakeholders where corporate profit is ultimately distributed among direct and indirect spending (material supply, sub-contractors spending and employee salary, etc.), national and local taxes and accumulated profit (usually to be distributed to shareholders). Therefore, subtracting local distribution can capture additional values induced through the creation of local enterprise (Suwa, 2020).

In this study, the supply chain analysis was applied to an individual local renewable energy corporation to analyze differences in the local economy. It particularly focuses on (1) municipal tax income, (2) disposal income of employees and (3) after-tax profit of the corporation to find out the degree of the regional contribution of such renewable corporation. Through the analysis, it reveals that profit and its associated value added would not remain local, unless the plant owners and their business partners are locally external players, as the tax income will be incurred elsewhere (Nakayama, et al., 2016).

The basis for calculating the effects obtained from local contributions was the addition of local taxes at prefecture and city levels, salaries to locally hired personnel and spending to regional renewable suppliers. The methodological justification is broadly based on Heinbach et al. (2014) that municipalities profit in two ways: enterprise profits are subject to local tax paid to the municipality (whereas corporate tax goes to the national government budget). In addition, in order to reflect the real business practice that MPSs purchase renewable energy from a regional supplier, their spending was identified as a component for supply chain contribution. Taken together, these could form benefits generated by the establishment and operation of an MPS.

This study analyzes two new municipal power suppliers in order to measure the level of economic circulation they generate in the local communities given each business structure. These two MPSs were chosen as cases because they have been pioneers among the Japanese MPSs, with a record of local stakeholder participation.

Case studies: new municipal power suppliers subject to case analysis

Miyama Smart Energy

Miyama city is in the southern part of Fukuoka prefecture, about 50 km south of Fukuoka city, Kyushu Island. Most of the city area is in the Tsukushi Plain (Tsukugo Plain), and the southwestern part of the city faces the Ariake Sea. The key industry is agriculture. The eastern side of the Kyushu Expressway, which runs north-south through the eastern part of the city.

As is often the case with other Japanese locality, it faces expected population decline (Figure 7.1), which means the city government is keen to secure revenues in various forms for maintaining public services.

Founded through joint investment by Miyama city in Fukuoka prefecture and local companies, Miyama Smart Energy procures electricity from surplus power generated by residential photovoltaic systems and large-scale solar power generation plants in Miyama city and mainly sells the electricity to the Miyama city government. It has a contract demand of approximately 36 MW, one of the highest among the new power retailers funded by local governments in Japan. The company has also launched a diverse range of initiatives to address problems facing the community, including an elderly watch service using the home energy management system (HEMS), a package discount for power and water bills and a mail-order business aimed at reinvigorating the city's shopping district. Investments are mainly from the region, and almost all the company's employees, some 50 in total, live in the city (based on information as of July 2017).

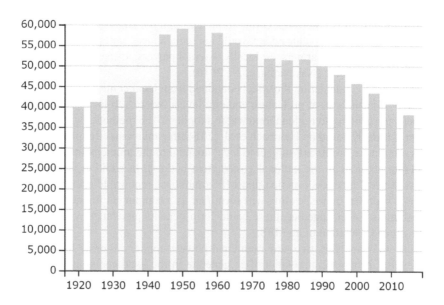

Figure 7.1 Population trend in Miyama city (Source: Miyama city homepage).

Value added to the local economy by Miyama Smart Energy

Key financial figures are taken from the company's business plan for 2017 to estimate the value added to the local economy by Miyama Smart Energy, which includes (1) operating income (approx. 21 million yen) and net profits (approx. 14 million yen) from sales (approx. 1.4 billion yen); (2) increases in the net profits of contractors to which Miyama Smart Energy has made payments and the increase in the disposable incomes of the contractor's employees, if the contractors are determined to be locally stationed (approx. 2 million yen; estimated through the corporate financial data); (3) profits made by local residents as a result of the higher-than-normal price at which Miyama Smart Energy purchases surplus power from residential photovoltaic power generation systems (approx. 5 million yen); (4) the disposable incomes of local employees (approx. 75 million yen); and (5) municipal taxes (approx. 7 million yen).

This chapter estimates the value added to the local economy by Miyama Smart Energy is approximately 100 million yen. To determine to what extent Miyama Smart Energy contributes to the local economy, calculations were made to obtain the value added to the local economy if local investment accounts for only 10%, supply and demand management and other operations are outsourced to contractors from outside the community and all employees live outside the local community. It was found that this business structure resulted in the outflow of capital from the community, and value added would be approximately 9 million yen, one-tenth of the case otherwise (Figures 7.2 and 7.3).

Hioki Energy Co., Ltd

Hioki city is a city located in the central part of Kagoshima prefecture, also on Kyushu Island. Compared with other rural cities in Japan, the city maintains its

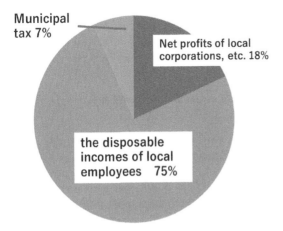

Figure 7.2 Breakdown of the value added to the local economy by Miyama Smart Energy.

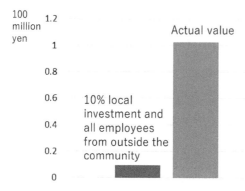

Figure 7.3 Alternative proportions of local investment and local staff lead to a significant difference in value added to the local economy.

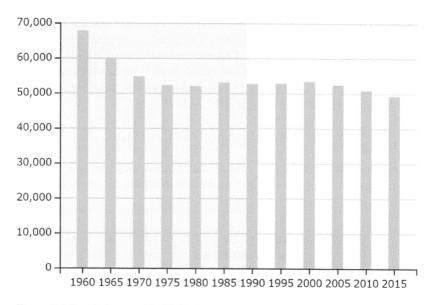

Figure 7.4 Population trend in Hioki city (Source: Hioki city homepage).

level of population (Figure 7.4), as it is close to Kagoshima city (a big city in the Kyushu area) and is developing as a suburban "bed" town.

Founded through joint investment by Hioki city in Kagoshima prefecture, local businesses and individual investors, Hioki Energy has a contract demand of approximately 7 MW. The company has three employees from the community and donates a portion of its earnings to the local community. The company is currently engaged in a small-scale local hydropower project (as of March 2017).

Value added to the local economy by Hioki Energy

Key financial figures are taken from the company's financial information in 2016 to estimate the value added to the local economy by Hioki Energy, which includes: (1) operating income (approx. 8 million yen) and net profits (approx. 6 million yen) from sales (approx. 140 million yen); (2) increases in the net profits of contractors to which Hioki Energy has made payments and in the disposable incomes of the contractor's employees, if the contractors are determined to be local based on interviews conducted with payees for each payment item (approx. 0.2 million yen; estimated using statistical data on private corporations); (3) the disposable incomes of local employees (approx. 3 million yen); and (4) municipal taxes (approx. 0.4 million yen).

The value added to the local economy by Hioki Energy is estimated at approximately 9 million yen. The same comparison with the Miyama case was made to obtain the value added to the local economy by assuming Hioki Energy had local investment of 10%, supply and demand management and other operations are outsourced to contractors from outside the community and all employees live outside the local community. It was found that this business structure resulted in the outflow of capital from the community in the forms of shareholder earnings and employee salaries; Hioki Energy's value added would be approximately 1 million yen, one-ninth of the case otherwise.

Furthermore, when additional analysis was conducted based on another scenario in which supply and demand management and other operations were performed within the local community, it was found that the value added to the local economy would be higher than the actual value, which is calculated based on the premise that these operations are outsourced to non-local corporations by approximately 40% (Figures 7.5 and 7.6).

Discussion and conclusion

In this study, the economic circulation generated by the operations of two new municipal power suppliers in Japan was measured by analyzing the value added to the local economies. The study results revealed that while one of the major goals that local governments seek to accomplish by establishing new power suppliers is to generate economic circulation in their local communities, the actual value added to the local economies varies by as much as nine to tenfold, depending on the business structure of the new power supplier. It was demonstrated that the means used to achieve the purpose of establishing a new power supplier does not always serve that end.

While one of the major goals local governments seek to accomplish by launching MPSs is to achieve economic circulation in the local community, the results of this study demonstrate that the means employed do not always accomplish the goal depending on the business structure of the new power retailer. If a new power supplier is established for the purpose of generating economic circulation in the local community, it is important to create a structure that allows operations to be performed within the community as much as possible.

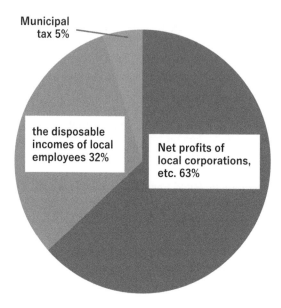

Figure 7.5 Breakdown of the value added to the local economy by Hioki Energy.

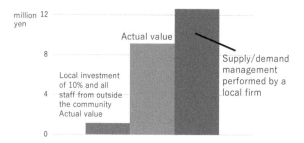

Figure 7.6 In-house performance of operations leads to a significant difference in value added to the local economy.

This finding makes it clear that simply establishing a new power company does not necessarily lead to an increase in value added to the local economy and that the choice of business structure is a key consideration that affects the value. Specifically, it is crucial to fund the new company with local capital and employ local staff to perform operations in-house to ensure gains for the local community.

As mentioned, there is a high hope that the Stadtwerke model – in which profits from energy operations are shared with the community as described earlier – will be applicable to Japan as the country seeks to maintain its urban infrastructure while battling low birth rates, aging and population decline.

It is, however, important to note that Japan's new municipal power suppliers do not have such solid financial revenue sources as some of their European counterparts. In Japan, the existing power transmission and distribution networks are owned by major power utilities. Even if a new municipal power supplier seeks to start a power project in a specified area, it needs to install new, costly power lines and operate them independently. Most new power retailers are therefore not entering the power distribution business, one of Stadtwerke's revenue sources, nor have they ventured into the district heating and cooling business, in which a large initial investment is required, and profitability is difficult to secure.

In addition, MPSs must develop strategies to respond to various risks in the power retailing business, their principal operation. These include securing independent power sources that are not impacted by wholesale electricity prices (by purchasing electricity jointly with other new municipal power suppliers from major public hydroelectric plants, for example) and effectively sharing information regarding power-related institutional changes directly related to management strategies through collaboration with other new municipal power suppliers.

In particular, regulatory risks faced by new power suppliers – many of which are small businesses – in managing supply and demand independently were significantly reduced when changes were made to the method of calculating imbalance rates as a result of the power market reforms in fiscal 2016. Until fiscal 2017, rates for any imbalance over 3% of the total demand exceeded 50 yen/kWh in some cases. As a result of the system reforms, the 3% threshold was abolished, and a new calculation method based on the market price was introduced. In addition, less costly supply and demand management systems with limited features are starting to appear on the market. These changes are gradually laying the foundation that allows the new power suppliers to manage power supply and demand in-house.

Once their business is on the right track, new municipal power suppliers can be expected to enter businesses that are synergistic with power retailing such as renewable energy development operations and play a role in enhancing economic circulation in local communities and lowering carbon emissions in the region. Compared with other regional revitalization projects, new municipal power retailing may have a higher probability of success. Expectations are running high that new municipal power suppliers across Japan will help reinvigorate local communities and lower their carbon emissions, and the national government surely has its responsibility to invigorate MPSs by establishing a business environment where fair and transparent corporate operation is secured.

Acknowledgments

This work was supported by JSPS KAKENHI Grant Number 15H01756 and The Toyota Foundation Grant Number D18-R-0133

Notes

1 The power retail market in Japan has been gradually liberalized with the deregulation of extra-high voltage (2000 kW or more) in March 2000, high voltage (500 kW or more) in April 2004 and high voltage (50 kW or more) in April 2005. Full market liberalization took effect in April 2016.

2 In Germany, the Energy Industry Act was enacted in 1998 to implement the full liberalization of the power retail market. Twenty years have since passed, and the Stadtwerke have been able to successfully compete with major power utilities such as RWE and E.ON. Reasons for the Stadtwerke's competitiveness include: (1) a branding strategy that aims to maximize citizens' life satisfaction, instead of maximizing profits; (2) services closely tied to the community; and (3) environmentally conscious customers acquired by utilizing renewable energy.

3 A mechanism in which multiple power retailers form a group and choose a representative, which then carries out supply and demand management and other operations for the entire group. By reducing the work of power retailers other than the representative and achieving a leveling effect by balancing the group's demand, the mechanism is expected to reduce "imbalance" fees (fees paid to general power utilities based on the difference between planned power and the actual power used).

References

Bulkeley, H., Broto, V.C. and Maasen, A. (2014) 'Low-carbon transitions and the reconfiguration of urban infrastructure', *Urban Studies*, 51(7), pp.1471–1486.

Hall, S., Foxton, T.J. and Bolton, R. (2016) 'Financing the civic energy sector: How financial institutions affect ownership models in Germany and the United Kingdom', *Energy Research & Social Science*, 12, pp. 5–15.

Heinbach, K., Aretzs, A. and Hirschl, B. (2014) 'Renewable energies and their impact on local value added and employment', *Energy, Sustainability and Society, Springer Open Journal*, https://link.springer.com/article/10.1186/2192-0567-4-1.

Jenniches, S. (2018) 'Assessing the regional economic impacts of renewable energy sources – A literature review', *Renewable and Sustainable Energy Reviews*, 93(2018), pp. 35–51.

Koirala, B.P., Koliou, E., Friege, J., Hakvoort, R.A. and Herder, P.M. (2016) 'Energetic communities for community energy: A review of key issues and trends shaping integrated community energy systems', *Renewable Sustainable Energy Review*, 56, pp. 722–744.

Morotomi, T. (2016) 'The establishment of municipal energy utilities and their significance', *Cities and Governance*, 26, pp. 59–70.

Morotomi, T. (2018) 'Mature urban management amid population decline', *Regional Finances*, Vol.2018.10, pp. 4–18.

Nakayama, T., Raupach, S.J. and Morotomi, T. (2016) 'Regional value added analysis of renewable energies in Japan: Verification and application of regional value added modelling in Japan', *Research on Sustainability*, 6, pp. 101–115.

Ogawa, Y. and Raupach-Sumiya, J. (2018) 'Economic effects of renewable energy to the community: A case study using value chain analysis', *Environmental Science*, 31(1), pp. 34–42.

Public Electric Utility Enterpriser's Forum (2018) 'Organizational profile', http://koueidenki.org/data/koueidenki.pdf

Raupach, S.J. (2014) 'Measuring regional economic value-added of renewable energy: The case of Germany', *Social Systems Studies*, 29, pp. 1–31.

Raupach, S.J., Nakayama, T. and Morotomi, T. (2015) 'Regional economic effects of renewable energy in Japan: A calculation model based on industry chain analysis specific for each power source', *Renewable Energy and Regional Regeneration*, Nippon Hyoron sya Co.,LTD. Publishers, pp. 125–146.

Stadtwerke Ettlingen Gmbh. Company Profile. https://www.buhlsche-muehle.de/fileadmin/pdf/SWE-Standortbericht.pdf

Suwa, A. (2020) 'Renewable energy and regional value: Identifying value added of public power producer and suppliers in Japan', *Finance Research Letters*, 37, p. 101365.

Yajima, M. (2017) 'Management strategies and issues of municipal companies in Germany: Focusing on the electricity business', *International Public Economy Studies*, 28, pp. 36–45.

Yamashita, H., Fujii, K. and Yamashita, N. (2018) 'A study on the actual conditions of renewable energy use in the Japanese municipalities', *Hitotsubashi Economics*, 11, pp. 175–221.

Part IV

Territories with 100% renewable energy

8 The voluntary initiatives, "positive energy territory" and "positive energy territory for green growth", first steps toward decentralization of the French energy system?

Blanche Lormeteau

In order to overcome the various forms of energy dependence[1] (Debeir et al., 2013; CGDD, 2018) and the climate emergency, the resilience of territories (Walker et al., 2004; Mathevet and Bousquet, 2014), that is, their capacity to change while maintaining their identity (Mathevet and Bousquet, 2014), is essential. This ability to change, to absorb natural and anthropogenic disturbances, is based on a territory's capacity to adapt and its ability to "make the transition", that is, to transform itself. In this sense, a resilient territory is essentially based on two principles: systemic interactions (high level of modularity and diversity of its components – ecological, social, economic) and solidarity (within the territory and with neighboring territories) (Mathevet, 2012; Michelot, 2018).

The search for resilience is also based on the capacity of institutions to develop adaptive management coupled with collaborative management (Ostrom, 2010). In this sense, resilience is based on public control of institutional tools (Aykut and Evrard, 2017), public governance, understood as a "pluralist and interactive or negotiated approach to collective action" (Chevallier, 2003), and even, for the audacious theories, by citizen self-management of resources (Lagneau, 2013). Consequently, the resilience of territories deals with the management of all the resources (energy, economic, food, social, health, etc.) to foster adaptation to climate change.

From this initial questioning emerges a new methodology for apprehending local territories, that is, the ecological transition as a search for the most effective and efficient level of decision-making to fight and adapt to the climate emergency (Van Lang, 2018; Goujon and Magnan, 2018). Who knows best the resources and needs of the territory? Who can best influence the use of resources and the satisfaction of needs? At the very *least*, applied to energy, what underpins the ecological transition is the control, by local actors, of the energy flows and stocks of a territory, by means of planning and development documents, through local production and decentralized energy distribution and local energy governance.

DOI: 10.4324/9781003025962-13

French local initiatives such as "positive energy territories" (in French *Territoire à energie positive,* TEPOS) and positive energy territories for green growth (*Territoire à energie positive pour la croissance verte, TEPCV*) are part of this search for territorial resilience.

These two initiatives respond to parallel stories. The TEPOS were born in the 1990s from the will of a group of Breton farmers confronted with the phenomenon of green algae. They realized that this resource could be used for energy purposes, which would reduce the energy imports of their territory, thus reducing the energy and therefore economic vulnerability of these rural territories and managing one of the environmental consequences of pig farming. They started carrying out methanation projects, which in the 1990s were only slightly developed or even unknown to local authorities (Riollet and Garabuau-Moussaoui, 2015).

Assessing the vulnerability of a territory is at the basis of the TEPOS approach and constitutes a first appropriation of the national energy objective at the local level, in this case, energy security.

As part of this search for energy security, the TEPOS approach is based first of all on the identification and integration of all the energy risks of a territory to guarantee its resilience. The approach is particularly new on this point and has a direct influence on the mode of governance of these territories. The following are thus identified as risks linked to the governance of energy, creating a vulnerability of the territory: industrial risks (accidents on production facilities – in particular, the nuclear industry and its waste, oil spills); climate risks (manifestations of climate disruption – extreme weather events, fires, loss of biodiversity, drop in agricultural production, etc.); health risks (air pollution and the consequences on health – premature death, chronic disease); and terrorist risks (cyber or physical attack) (TEPOS, institutional website; Nadaï et al., 2015).

Another general characteristic is that the TEPOS approach is neither regulatory nor standardized. It is a label, a brand, registered by CLER (Network for Energy Transition). No methodological tool is specifically associated with the TEPOS approach. Not being regulatory and corresponding to the delivery of a label, all types of territories can claim to obtain this label, such as intermunicipal groupings, *régions*, the Regional Natural Parks and so on, as long as they adhere to the main principles of governance of the label and that they carry ambitious energy objectives.

The TEPOS are gradually being integrated into the institutional landscape. This can be seen in the "TEPOS" earmarked funding, developed and supported by *régions* and public central institutions (Caisse des dépôts et consignation, ADEME, etc.). A minor legal recognition since the law on Energy Transition for Green Growth (Law ETGG) of 2015[2] is another sign of this recognition. It is therefore through funding that the TEPOS became visible at the national level, and this came with a certain recognition of their effectiveness.

Yet, wishing to rely on territories to meet the objectives of the national energy policy in the context of COP 21, the French legislator enshrined in law the concepts of "positive energy territory for green growth" (TEPCV). This new label corresponds to a program of actions labeled and financed by the state following a national call for events and is regularly confused with the TEPOS because they sound alike.

This interest in territories and the voluntarist aspect of these two initiatives are part of a national policy, which raises questions about the effectiveness and room for maneuver available to them to achieve the objectives of reducing energy needs as much as possible, through energy conservation (or in French "*sobriété*") and efficiency, and to cover them with local renewable energies ("100% renewable and more").

Indeed, these two initiatives are born in a very particular context, that of a deconcentration, or even a functional decentralization,[3] of energy and climate policies in France. Historically centralized and dominated by the sovereign power of the state, the French energy system has been organized around large production and distribution facilities[4] (Debeir et al., 2013; Poupeau, 2007; Lopez, 2019). The economic actors, few in number, were then subject to the structural choices of the state, which favored the forms and sources of energy allowing continuous production[5] (Puiseux, 1978). However, the development of renewable energy sources, available in all areas (Petit, 2016), and the increasing introduction of issues relating to the fight against and adaptation to climate change are giving territories new challenges to take up (Marcou et al., 2015), which has resulted, under French law, in an increase in the competence of local players, with local authorities in the lead.[6] Although a certain transfer of energy competences has been institutionalized, from the central state to local authorities, this contribution aims to answer the question of whether these local initiatives, the labeling of a territory, are also the marker of an even more effective decentralization of the French energy system?

The answer is positive. The TEPOSs are based on a radically innovative methodology, fully in line with the search for territorial resilience. Far from meeting strict specifications, this label corresponds to recovering local energy governance according to the needs, capacities and resources of the territory. Although this new type of governance is not welcomed by the state, the TEPOSs are at the heart of institutional and conceptual experiments that have been partially taken up by the legislature and the executive.

However, observation also shows the state's refusal to allow the territories to completely take over energy issues, as the example of the methodology implemented within the framework of the TEPCV and the massive recourse to contractualization show.

Innovative rules of governance are not easily absorbed by the central level

The TEPOSs, in addition to the fact that they are mainly rural territories, are based on the will to create a virtuous local loop (TEPOS, institutional website), especially because the award of the label does not come with funding. So first they aim at a reduction in energy needs (energy conservation and efficiency), which makes it possible to generate savings. Second, thanks to the savings, they finance energy efficiency actions and the development of local renewable energy production plants so that all remaining needs are covered in the long term by local renewable energy sources (in production and import). In principle, it is through the reduction of needs that self-financing is created. However, of course, there is nothing to prevent the territories concerned from seeking sources of funding to initiate virtuous approaches, creating the link with the funding promised by the TEPCV label.

From this virtuous loop, several guiding principles of the TEPOS governance are derived, which consist in the integration of the search for territorial resilience: energy neutrality; self-sufficiency; energy autonomy; "100% renewable energies" territory; territorial solidarity; and local energy democracy (TEPOS, institutional website).

While there is no single set of instructions for obtaining the TEPOS label, it is possible to identify four markers of recovery by the local level of energy governance: the notion of active subsidiarity; reasoning by flow; the strengthening of territorial engineering and the development of a network; and energy democracy.

The notion of "active subsidiarity" guarantees the territories freedom to choose the objectives to be achieved according to local potential and culture. These local objectives contribute to the satisfaction of national objectives, but local authorities determine them in line with national policies. From then on, the TEPOS takes over the existing tools and develops a real command of them. This way, the issues of home energy are integrated with those of food for people and livestock, and mobility, thus enabling them to work on all the factors of energy vulnerability. Therefore, legally defined planning tools such as PCAETs (Angot and Gabillet, 2015), and Agenda 21, whose normativity is weak (Lormeteau, 2020), become unavoidable instruments of the TEPOS approach by participating in the initial diagnosis and in the determination of objectives and actions very specific to the territory without being a mere variation of the national objectives. Moreover, TEPOS really questions the usually very technical approach of energy (production and distribution installations) to the benefit of a more flow-based methodology.

Flow reasoning is a particular methodology that allows the resilient management of a territory. TEPOSs do not develop an action program aimed at achieving, in an accounting way, an equal output between the production of renewable and local energy and consumption, which is more in line with national policy (SNBC, 2020). Rather, the TEPOS works on the energy flow of its territory, that is, it includes in its reasoning the exported productions and the imported consumptions. Thus, energy is not limited to the forms of energy produced and consumed, but also to the goods and services produced or imported on its territory. It is therefore an approach based on the territorial carbon footprint that can be developed dynamically.[7] It includes: "Scope 1" emissions, generated by the production of goods and services (combustion sources, biomass); "Scope 2" emissions, necessary for this production (consumption of electricity or steam, heat and cold); and "Scope 3" emissions, related to activities upstream of production (purchase of products and services, upstream transport of goods, business travel).

This understanding of energy is gradually influencing national policies, taking into account the experimental role of the TEPOS and their progressive institution-alization. One example is the new obligations of companies with regard to their social responsibility, which implies an obligation to provide environmental infor-mation[8] on the climate impact of the goods and services they produce (including abroad) (Mabile and de Cambiaire, 2019).

However, this institutionalization of the flow methodology remains very par-tial, as the integration of the experiments into ordinary law shows (*infra.*). There is an individualization of the energy issue rather than a synergy between all activi-ties that can have an impact on the reduction of energy needs.

This is peculiar to French energy law, which remains a sectoral legal branch, piloted and determined at the central level (Terneyre and Boiteau, 2017; Sablière, 2016). It does not allow for a flow methodology. Thus, energy planning documents (SRADDET,[9] SRCAE,[10] PCAET, etc.) are still largely based on the satisfaction of national objectives that are not sufficiently broken down according to local charac-teristics and are still based on purely the accounting method, the effectiveness and efficiency of which is difficult to quantify[11] (Poupeau, 2013; CEREMA, 2018).

The reinforcement of territorial engineering and the development of a net-work is the third marker of the TEPOS methodology. The TEPOS governance is not only organized at the scale of the labeled territory but at the national level, through a network used as a basis for another pillar of these territories: training.

In fact, in order to take over all energy issues, particular attention is paid to the skill development of elected representatives and territorial agents by the con-stitution of real territorial engineering. Several modalities are promoted, such as the creation of a position of mission manager and the regular training of elected representatives and agents in energy issues (Nadaï et al., 2015). For example, the call for projects "TEPOS" of the Nouvelle Aquitaine region in 2017 included a proposal for project management assistance (AMO). It focused on the structur-ing of local energy governance to support the training of elected officials and the systematic introduction of energy-climate issues during the vote on budgets. It also provided for a technical AMO on the technical and legal training of elected officials on the conduct of renewable energy projects.

This training is combined with the pooling of tools for modeling energy choices and the provision of technical and technological support through a national and European network. The TEPOSs are members of the "100% RES community" network,[12] which federates the different initiatives carried out by territories throughout Europe in terms of energy transition. It aims at sharing experience through exchanges and study trips and contributes to developing a methodology through the definition of common indicators. It also undertakes lobbying within the framework of the Covenant of Mayors.

The fourth and final key marker is that of "energy citizenship" (Calandri, 2015), that is, involving citizens in determining energy choices and participatory financing, understood in a broad sense. This mode of citizen participation was a major innovation of the TEPOS and has since been widely taken up by the legisla-tor, acting on the state vision of the TEPOS as a field of experimentation.

TEPOS: fruitful experimentation grounds for a hesitant state

Local authorities and their groupings in France have a right to experimentation. This is allowed in specific situations (Stahl, 2010). The first arises from Article 37-1 of the Constitution, which provides that "laws and regulations may contain provisions of an experimental nature for a limited purpose and duration"; the second arises from the application of paragraph 4 of Article 72 of the Constitution (Fialaire, 2004; Conseil d'état, 2019), which allows local authorities to depart from the limitation of their powers by means of express legislative and regulatory provisions;[13] the third results from more factual situations, when authorities with normative power make experiments within their jurisdiction, but without specific legislative or regulatory powers.[14]

In the field of energy, as previously presented,[15] the territories have few direct responsibilities. They, therefore, develop experiments in other sectors, related to a block of local climate-air-energy policies, for which they are responsible since the adoption of the Grenelle laws in 2009 and 2010 (regional planning, development of local food and energy production sectors, energy performance of housing, mobility, etc.). At a later stage, the legislator and the executive power can validate their initiatives.

The first innovation is the very conceptualization, as of 2011, of the concept of "positive energy territory", subsequently institutionalized in 2015 by the Law ETGG. Its first article states that:

> A positive energy territory is a territory that commits itself to a process that makes it possible to achieve a balance between energy consumption and production at the local level by reducing energy needs as much as possible and respecting the balance of national energy systems. A positive energy territory must promote energy efficiency, the reduction of greenhouse gas emissions and the reduction of the consumption of fossil fuels and aim at the deployment of renewable energies in its supply.[16]

Focused on energy in the strict sense of the term (production and consumption), this definition has the merit of integrating in law the concept of a "positive energy territory" but it omits a key feature: the methodology of designing local public policies through flow. In this sense, the definition only partially takes up the methodology of resilience that drives the TEPOS, that is, the integration of all ecosystem resources. However, it is on the basis of this definition that policies related to TEPCVs will be developed. Consequently, the legislator takes over a label created by local and regional authorities to specify a methodology of action, modifies its substance and makes it the reason for a new contractualization – between the state and the authorities – of public action. The territory is no longer the driving force, it is the state.

Another conceptual experimentation carried out by the TEPOS and then institutionalized by the legislator is the integration of the objective of carbon neutrality as a basis for local policies.

Adopted in 2007 by one of the pioneering TEPOS territories[17] (CLER, 2010), carbon neutrality is now one of the guiding objectives of French energy law since the adoption of the 2019 Energy and Climate Law.[18] If the international context also contributes to this legislative recognition,[19] it is however interesting to note that the TEPOS territories had already used this concept to build their energy policy. It has been the basis of local clean energy and climate objectives, but also of the new tools, such as local climate currencies. Thus, the local policy of the La Rochelle conurbation community, on the basis of a local consortium,[20] developed "La Rochelle Territoire Zéro Carbone" at the end of April 2019. Drawing on its experience with TEPOS to carry out an innovative project redefining the land/ocean relationship, a policy of carbon offsetting has emerged as the preferred way to measure and then limit man's impact on the environment and preserve the quality of life. Offsetting is part of a local approach, with each emission from the territory having to be offset on the territory and is similar to a local currency.[21] The consortium from La Rochelle thus aims to reduce the territory's carbon footprint by 30% by 2030 and to achieve complete carbon offsetting by 2040 in order to propose a virtuous model that can be replicated in other territories.

Another experiment that is now integrated, or even promoted in law, is the participatory financing and cooperative governance of projects for the production, and now consumption, of local renewable energy. Thus, from 2011, some TEPOSs include citizens (consumers or not) and a public body (consumer or not) in the capital of heat production installations[22] (Lormeteau, 2014), then in the capital of wind and solar projects (Allemand and Dreyfus, 2017). Since the Law ETGG, the legislator has constantly opened up the possibilities for citizens and local authorities to enter into the capital of private companies producing renewable and local energy or to create their own public company (Fontenelle de, 2019). It has also created a dedicated label[23] and, above all, has accepted the notions of collective self-consumption[24] and a renewable energy community,[25] which implies local production and consumption of the energy produced (Lormeteau and Molinéro, 2018).

The TEPOSs also carry out more technological experiments. For example, based on their skills in land use and urban planning, many TEPOSs have experimented with the implementation of a solar cadastre.[26] This tool provides information on the solar potential of buildings, and the energy potential, and therefore the economic profitability of a solar thermal or photovoltaic system. They are now one of the irrefutable conditions for obtaining the "Solar City" and "Solar Department" labels created as part of the government's "Place Au Soleil" plan launched in June 2018; 100 territories should be labeled by 2020.[27]

However, the real experimentation carried out is that of the governance of the TEPOS, which is only very partially acknowledged by the state, proof that the experiments carried out by the TEPOS do not lead to a real acceleration of the energy transition.

This governance by the flow, the increase in competence of elected representatives and territorial agents, as well as the integration of citizens in the decision-making process of the TEPOS, show that local authorities aspire to take over in the long term the issue of energy transition and the transformation of their territory into a resilient territory. The attempts of the state to capture the territorial dynamics through TEPCVs, and now energy transition contracts, are moving away from this innovative methodology, revealing the refusal of the state to proceed to an effective decentralization of energy governance and favoring a top-down energy territorialization alongside territory projects (Durand et al., 2015; Bailleul, 2019).

The state's mistrust and the failure of the TEPCVs

The introduction by the Law ETGG of the concept of "positive energy territory" is coupled with a first attempt to institutionalize the TEPOS by creating a national and regulatory label: the TEPCV.

TEPCVs are defined as an action program focusing on reducing the energy needs of its inhabitants, buildings, economic activities, transport and leisure activities. Although their scope is similar to that covered by the TEPOS, except that TEPCVs focus on energy issues, their implementation method is fundamentally different. Indeed, in that case, the central government keeps under scrutiny local initiatives through a regulatory and contractual framework. In fact, the contractual technique is traditionally used to integrate, if not impose, national energy objectives at the subnational level.

To this end, the TEPCVs identify six priority areas for action: reducing energy consumption; reducing pollution and developing clean transport; developing renewable energies; preserving biodiversity; preventing and reducing waste; and environmental education.

All of these priority actions *ultimately* correspond to the application of existing law, in terms of its principles, tools and methods, and objectives. Consequently, the future contractualization is indeed a *top-down* approach and not a *bottom-up* one as in the case of the TEPOS. Through the funding granted by the TEPCV label, the state imposes and funds the effectiveness of national provisions and objectives at the local level; it decides which ones will be favored according to dedicated and time-limited calls for projects. As C. Guettier points out, contractualization makes it possible

> to obtain the active collaboration of peripheral units in the implementation of priorities set by the central level, without the latter having to resort to coercion: in order to obtain financial assistance from the State, peripheral units are led to adhere to its rationality, to internalise its standards and to take over its objectives on their own account. (Guettier, 2005)

Another characteristic of this top-down approach, to respond to the six priority areas of action, is that territories are labeled following a national call for projects, the criteria of which are not determined to respond to a local need, but in a general manner, with a view to adding up the results obtained in order to meet the objectives of the national energy policy. Thus, contrary to the TEPOS, which focus on the implementation of a new methodology in energy governance,

TEPCVs are built around actions to be carried out in order to participate in "green growth", which seems antinomic with respect to the principle of energy conservation ("*sobriété*") defended by the TEPOS (Audrain-Demey, 2018).

Similarly, prerequisites were required: the existence of local engineering, a co-built territory project and as options: direct or delegated operational skills in terms of local ecological transition (transport, energy distribution, etc.). The prerequisite of the existence of local engineering capacity highlights the fundamental difference between TEPCV and TEPOS: consisting in financing actions to be carried out to satisfy long-term objectives, the accounting logic prevails in TEPCV rather than a long-term construction of local energy governance allowing the territory to deploy a resilient and adaptive policy to the evolution of its specific needs.[28]

The proposals received for the first call for projects, covering the 2014–2016 period, unveil the state's determination not to support territories in a genuine decentralization of energy governance, confining itself to financing "exemplary" and experimental actions. The call thus aimed to "Mitigate the effects of climate change and present 'exemplary' territories at COP 21"; "Encourage the reduction of energy needs and the development of local renewable energies"; "Facilitate the establishment of green industries to create 100,000 jobs over 3 years"; "Reclaim biodiversity and enhance the value of natural heritage"; "Show the leading role of territories in the energy transition"; "Create territorial dynamics".[29] From a factual point of view, the call for initiatives was launched on 4 September 2014, 355 territories were TEPCV winners by 15 September 2016 in mainland France and overseas territories. The aid varied from €500,000 to €2,000,000 per winner (it comes from the €1.5 billion [over three years] energy transition financing fund). The labeled territories mainly develop targeted actions that are not necessarily correlated by a transversal vision of the energy issue. Thus, there is no global vision through the energy flow but on the objectives to be reached, and this is reflected in the actions proposed and financed. The majority correspond to actions of energy renovation and exemplary construction on the public heritage, followed by action on clean mobility, modernization of public lighting and the production of renewable energy. On the other hand, no participatory financing or civic service projects on energy transition have been carried out, and very few on the fight against waste, industrial ecology or nature in cities (B&L évolution, 2017).

However, it should be noted that this call for projects for TEPCVs has enabled many TEPOS to obtain funding for the actions undertaken, without this funding necessarily influencing the governance and methodology adopted by the TEPOS label, as the two labels are not exclusive (CLER, 2017). Thus, and this is a short-term limitation of the TEPOS, the virtuous loop that should allow self-financing is not necessarily feasible in all territories. If it seems possible in small rural territories, however, it is more difficult to be effective in peripheral or urban territories, corresponding to a certain level of population density. Therefore, if territorialization is taking place, the state continues to support the approach through the financial support of TEPCV to TEPOS.

Yet, the TEPVCs are deemed to have failed, and the program is not renewed. During the debate on the 2019 Public Finance Law (PF Law), the government refused to open a new call for proposals for the TEPCVs. The content of the exchanges reveals a certain mistrust of the central state toward the territories, and more generally its willingness to change the method of financing local energy initiatives.

In this sense, via an amendment to PF Law,[30] "sustainable development and mobility" mission credits, some members of parliament (MP) wished to maintain the TEPCV mechanism by allocating part of the budget for this budgetary mission to the financing of "new generation" TEPCVs, and to that end, a "special fund for the ecological transition of the territories" would be created. C. Bouillon, MP, defended the amendment, emphasizing the role of local authorities:

> Trust the municipalities, support them, give them the opportunity to revitalize positive energy territories [...] You will agree with me, Mr. Minister of State, that the success of the energy transition depends on the territories, i.e. on the mobilization of citizens and the emergence of projects in the fields of renewable energies, biodiversity and environmental education.[31]

As recommended by the principles of resilience and adaptability, which places territories at the heart of the success of the energy transition, the government did not seem to support this approach. Moreover, the response of the Minister of Ecology, Mr. de Rugy, reveals the central government's reluctance toward local's responsibility of energy policy:

> Many local authorities are obviously taking initiatives in favor of the ecological transition, whether in the field of energy, transport or water. There is a simple reason for this: it is at the heart of their responsibilities as far as municipalities or inter-municipalities are concerned. [...] You know local authorities and territories well, and you know that we can always ask ourselves: was there a leverage effect – which allowed actions that would not have taken place otherwise – or a windfall effect – the financing of actions that would have taken place anyway? This is why the government favors policies that support actions that are sure to have a leverage effect.

While the competences of local authorities are highlighted, it is the management of the financing of the actions that is criticized. This was taken over by the central state, even though, as the Court of Auditors pointed out, the difficulty does not come from the winners, but comes in particular from legal risks created by the state due to the combination of several state financial arrangements and the lack of dedicated funding provided for them (Court of Auditors, 2017).

The creation of ETCs: the end of a territorial approach to energy?

Based on a highly sector-based definition of the concept of "positive energy territory", the TEPCV label developed by the state failed. It was replaced by a more usual contractual logic for public energy action, the energy transition contracts (ETC). This new tool puts aside the idea of energy decentralization in favor of the state control of energy territorialization, which goes against the spirit of the resilience methodology.

The ETC mechanism is not intended to respond to a local energy governance issue and is an act of a reinforcement of the contractualization of the national energy policy combined with a certain withdrawal of the central state to the benefit of private actors and operators. However, this disengagement is only financial; the balance of local energy policy always responds to the same "recipe [...]: that of a dish whose content and flavor is determined solely by the State, in the name of its unitary character" (Kada, 2019).

The ETCs, like the TEPCVs, are the result of a national call for projects (in 2018 and 2019). They have two characteristics: they rally around all the stakeholders in the ecological transition, particularly associations and businesses – thus, perpetuating an actor-driven approach; they are built on local solutions, supported by stakeholders in the field. The Minister of State for Ecology, E. Wargon, stated that "The ecological transition contract illustrates the method desired by the government to support the territories: a co-construction with elected representatives, businesses and citizens who are betting on an ecological transition that will generate economic activities and social opportunities".

The objectives assigned to them are as follows:

> 1) Demonstrate through action that ecology is a driving force of the economy, and develop local employment through ecological transition (structuring of sectors, creation of training courses); 2) Act with all the actors in the area, both public and private, to give concrete expression to ecological transition; 3) Provide operational support in situations of industrial conversion of an area (vocational training, site conversion).[32]

The stated objectives seem to discover the role of the territory. The state again promotes an operational and not institutional approach. It does not commit itself financially. The contractualization allows, for him, on the one hand, to promote the emergence of public/private financing, and on the other hand, to allow the realization of "concrete" actions in an accounting logic opposed to the governance by the flows of TEPOS. In this sense, the contracts are evolutionary: they can integrate new actions as the initial actions are carried out. Consequently, the "vector of interventionist policy" in terms of ecological transition remains a tripartite contract, under the control of the state (Kalflèche, 2018).

Territorialization therefore no longer involves public actors but private funding, the "local" character of which is neither required nor encouraged. With regard to the first contracts concluded, it is observed that the large private, national or international companies are the first partners, far from favoring the emergence of a local sector in this sense.

In the same way, if some of the contracts concluded include training actions,[33] in conformity with one of the key features of the TEPOS methodology, it should be underlined that they are not part of an interactional dynamic of the resilient management of a territory. In this sense, they remain embedded in their sector by intervention themes: agriculture, industry, energy production, demand management and so on, and the actions currently presented in the contracts are more akin to the financing of specific projects than to a real local governance of energy flows.

Conclusion

The two territorial initiatives, the TEPOS and the TEPCV, and in the future the CTE, are linked together but are not part of the same approach.

Although the two respond to different governance logics, each shows the importance given to the territory to meet the objectives of the national energy policy. The invocation of such a territorial force is indicative of a profound movement toward decentralization of energy management.

However, on the issue of decentralization of governance, the conclusion is different. Thus, the TEPOS network claims a territorialized, social and solidarity approach to energy issues, whereas the TEPCV is organized around issues of promoting exemplary actions and supporting "green" public procurement.

Thus, if they promote a long-term vision on energy in the territories: "more sober and more economical" for TEPCV, "100% renewable and more" for TEPOS; only the TEPOS can really territorialize the energy issue, thanks to a governance based on the control of local engineering. The TEPCVs are more exemplary *one-shot* actions.

In the same way, TEPOS and TEPCV are both promoting dialogue between local authorities and other actors. However, only the TEPOS is based on a network of territories working together, whereas TEPCVs are managed by the ministry, which created a "club" and managed a platform of presentation of the actions – a methodology moreover taken up for the CTEs.

Similarly, both initiatives take energy out of its usual analytical framework, the environment, to place it in a systemic dimension and promote the economic and social benefits of the energy and ecological transition. But where the TEPOSs develop a "local development" approach, the TEPCVs call for "green growth". However, green growth is based more on an accounting and market logic, and not on a flow analysis based more broadly on the incorporation of all internalities and externalities – including social ones – likely to develop a real resilience of territories to the challenges of fighting and adapting to the climate emergency.

Therefore, the TEPOSs are formidable vectors of innovation and technical and institutional experimentation and are in this sense true markers of the decentralization of energy management. However, because the objectives of energy policy are governed by law and because the funding provided by the state only covers sectoral actions, and not considering the transformation and ecological transition of territories by integrating resilience as a guiding principle of local policies, the decentralization of energy governance remains partial. However, any transition, particularly in the energy sector, means a readjustment of the democratic fabric, which has both material and cultural implications (Rumpala, 2015). While the former seems to find an echo in centrally led systems, the cultural transformation required for resilience through decentralization of the energy transition has not yet been achieved.

Notes

1 Understood as an extreme dependency that affects the autonomy of a community. The first form of dependence is ecological, i.e., a dependence on exhaustible fossil sources whose production, transformation and distribution methods generate GHGs. It can also be observed through the destruction of ecosystems, the search for fossil energy sources compromises their capacity to ensure all ecological functions and ecosystem services. The second form of dependence is geopolitical. The energy system is highly dependent on the stability of the places where fossil fuels are captured and distributed. The third form is economic, revealed by the energy intensity rate corresponding to the energy dependency ratio, i.e., the ratio between the country's energy consumption and its gross domestic product. Energy dependence is also recognizable by the lack of diversity of the preferred forms of energy, creating technical and technological dependence; France's energy intensity was 46.4% in 2018.

2 Art. 1, Law No. 2015-992 of 17 August 2015 on energy transition for green growth.

3 "Deconcentration" is a French technique for administering territory, allowing the state to exercise its authority from the center to the local constituencies within which the decentralized services responsible for representing it are located; "functional decentralization" means a transfer of powers from the state to a public authority with specialized competence.

4 In France, the electricity, gas and oil sectors are organized around transport routes developed uniformly thanks to the unification of the regions toward a central state.

5 The 1974 Messmer Plan in the French nuclear sector.

6 See Chapters 1 and 3.

7 As recommended in a very innovative way by ADEME: https://www.bilans-ges.ademe.fr/fr/accueil/contenu/index/page/Bilan%2BGES%2BTerritoires/siGras/0

8 Art. L. 225-102-1 of the Commercial Code.

9 Art. L. 4251-2 of the CGCT.

10 Art. L. 222-1 of the Environmental Code.

11 The SRADDET takes into account the National Low Carbon Strategy, it is true that the SRADDET, because of this simple report, seems to be a document that intensifies the decentralization of energy-climate policies; the SRCAE is an orientation and objective document, so it has no direct effect on individuals. In particular, the scheme sets out the regional guidelines at the level of the regional territory and by 2020 and 2050 for mitigating and adapting to the effects of climate change, in accordance with the national objectives set out in the Energy Code. The objectives of the PCAETs are compatible with the guidelines of the SRCAE, and therefore, compatible with the national objectives.

It should be noted, however, that ADEME has taken over and made available more comprehensive calculation tools and methodology, see, in particular, the single climate, air and energy reference system resulting from the merger of the Cit'ergie and Climat Pratic tools, combining ambitious levels of local objectives and actions determined according to a thematic, cross-cutting and progressive methodology (https://www.territoires-climat.ademe.fr/).

12 http://www.100-res-communities.eu/

13 Art. L.O. 1113-2 of the CGCT.

14 CE, 15 June 2007, *Centre d'éducation routière Gargan gare et autres*, n° 284773.

15 See Chapters 1 and 3.

16 Art. L. 100-2 of the Energy Code.
17 Thouarsais Community of Communes, within the framework of the Local Climate Initiative Contract concluded between the territory, ADEME and the region.
18 Art. 1, Law No. 2019-1147 of 8 November 2019 on energy and climate; Art. L. 100-4 of the Energy Code.
19 Art. 4 of the Paris Agreement; IPCC, *Global Warming of 1.5°C,* 2018; European Commission, *A Clean Planet for All: A Strategic Long-Term European Vision for a prosperous, modern, competitive and climate-neutral economy,* COM/2018/773 final.
20 The Urban Community, the City of La Rochelle, the University, Altantech, Port Atlantique and 130 other private partners (https://www.agglo-larochelle.fr/).
21 See also the Community of Thouarsais Municipalities, see https://www.thouars-communaute.fr/compte-CO2.
22 As thermal energy is not transported over long distances, its management by a heating network is traditionally a local competence, thus demonstrating the willingness of territories to make full use of the competences allocated by law.
23 https://www.ecologique-solidaire.gouv.fr/label-financement-participatif.
24 Art. L. 315-2 of the Energy Code.
25 Art. L. 211-3-2 of the Energy Code.
26 v. the proceedings of the 7th national meeting on *Energy and rural territories, towards positive energy territories,* Grand Figeac, 28 September 2017 (http://www.territoires-energie-positive.fr/echanger/rencontres-nationales/rencontres-nationales-2017/deploiement-du-solaire-generaliser-le-developpement-du-solaire-sur-un-territoire-identifier-et-equiper-les-surfaces-propices); for an example, see the solar cadastre of the communities of communes and the Terres de Lorraine country (https://www.terresdelorraine.org/fr/cadastre-solaire.html).
27 At the time of writing, no progress report on the implementation of this Plan has been provided by the government.
28 Note that the instruction of 26 May 2015 on the implementation of special agreements for positive energy territories for green growth, NOR: DEVK1511837J, BO min. Envir. n°2015/10, 10 June 2015 simply provides for the commitment of the winner "a) to designate a referent elected representative who will be the guarantor of the approach; b) to set up a project team led by a project manager at the territory level".
29 *Call for projects for positive energy territories for green growth,* 4 September 2014.
30 Amendment n°II-243, FDP for 2019 (n° 1255).
31 AN, XVth Legislature, Ordinary Session of 2018–2019, Full Report, First sitting of Monday 5 November 2018, discussion of the second part of the draft budget bill for 2019 (Nos 1255, 1302).
32 https://cte.ecologique-solidaire.gouv.fr/
33 See the contract concluded by the territory of the Haute Côte-d'Or (https://www.actu-environnement.com/ae/news/quatrieme-contrat-transition-ecologique-agriculture-32591.php4).

References

Allemand, R. and Dreyfus, M. (2017) 'Les collectivités territoriales et l'énergie à la lumière de la loi transition énergétique pour la croissance verte', *Mélanges en l'honneur du professeur Gérard Marcou,* IRJS Éditions, Paris, pp. 55–67.
Angot, S. and Gabillet, P. (2015) 'Pour une sociologie de la gouvernance politico-administrative interne des questions d'énergie-climat', in Zélem, M.-C. and Beslay, C. (eds) *Sociologie de l'énergie. Gouvernance et pratiques sociales.* Paris: CNRS ed. 2015, pp. 117–124.

Audrain-Demey, G. (2018) 'Une méthode contemporaine pour un concept vieillissant : transition écologique et développement durable', in Van Lang, A. (ed.) *Penser et mettre en oeuvre les transitions écologiques*. Paris: Mare et Martin, pp. 63–75.

Aykut, S. and Evrard, A. (2017) 'Transitions énergétiques et changements politiques', *International Journal of Comparative Politics*, 24(1–2), [online].

B&L évolution (2017) *TEPCV: Effet de levier ou rendez-vous manqué ?*, [online]. Paris: This study, carried out by a private firm, is to our knowledge the only freely accessible assessment of the TEPCV experimentation.

Bailleul, E. (2019) 'Le territoire et ses acteurs, fragile pilier de la transition énergétique française', Revue internationale et stratégique, 113(1), pp. 107–117.

Calandri, L. (2015) 'Les citoyens dans la gouvernance énergétique: "libre choix", "débat public"', in Marcou, G., Eiller, A.-C., Poupeau, F.-M. and Staropoli, C. (eds) *Governance and innovations in the energy system. New challenges for local and regional authorities?* Paris: L'Harmattan, pp. 151–170.

CEREMA (2018) *Bilan national des SRCAE - rapport pour la DGEC*, Paris. [online].

CGDD (2018) *Bilan énergétique de la France pour 2018*, Paris. [online].

Chevallier, J. (2003) 'La gouvernance, un nouveau paradigme étatique?', *RFAP*, 1, p. 203.

CLER (2010) 'Vers des territoires à énergie positive', *Synthèse de l'Assemblée Générale*, Lyon, 6–28 May 2010.

CLER (2017) *Pratiques méthodologiques des territoires à énergie positive*, Paris. [online].

Conseil d'état (2019) *Les expérimentations: comment innover dans la conduite des politiques publiques?*, Les études du Conseil d'état, Paris. [online].

Court of Auditors (2017) *Note d'analyse de l'exécution budgétaire*, Paris. [online]: "Funding for territories with positive energy for green growth (TEPCV) for the benefit of local and regional authorities represents 87% of cumulative commitments and 62% of cumulative payments. It also represents 95% of the outstanding balance."

Debeir, J.-C., Deléage, J.-P. and Hémery, D. (2013) *Une histoire de l'énergie: les servitudes de la puissance*, Flammarion, coll. Paris: Nouvelle bibliothèque scientifique.

Durand, L., Pecqueur, B. and Senil, N. (2015) 'La transition énergétique par la territorialisation, l'énergie comme ressource territoriale', in Scarwell, H.-J., Groux, A. and Leducq, D. (eds) *Transitions énergétiques: quelles dynamiques de changement?* Paris: L'Harmattan, Colloques et rencontres, pp. 211–222.

Fialaire, J. (2004) 'Le droit à l' expérimentation des collectivités territoriales et la subsidiarité: les apparences et les 'faux-semblants' d'une prétendue territorialisation des normes', in Fialaire, J. (ed.) *Subsidiarité infranationale et territorialisation des normes*. Rennes: PUR, p. 11.

Fontenelle de, L. (2019) 'Les communautés énergétiques', *EEI*, 8–9, p. 29.

Goujon, M. and Magnan, A. (2018) 'Apprehender la vulnérabilité au changement climatique, du local au global. Regards croisés', Paris. Ferdi-Iddri working paper, P215.

Guettier, C. (2005) 'Quel avenir pour les contrats de plan Etat-régions?', *RLCT*, 4, p. 30.

Kada, N. (2019) 'Etat et collectivités territoriales: (petite) cuisine et (grandes) dépendances', *AJDA*, pp. 2423–2429.

Kalflèche, G. (2018) 'Contractualisation et interventionnisme économique', *RFDA*, 2, p. 214.

Lagneau, A. (2013) 'Ecologie sociale et transition. Interview avec Vincent Gerber', *Mouvements*, 3, p. 77.

Lopez, F. (2019) *L'ordre électrique, Infrastructures énergétiques et territoires*. Genève: METISPRESSES.

Lormeteau, B. (2014) *L'énergie thermique et son droit*, thesis, University of Nantes.

Lormeteau, B. (2020) 'Contentieux nationaux de la planification climatique', in Torre-Schaub, M. (ed.) Lormeteau B. (coll.), *Les contentieux climatiques. Dynamiques en France et dans le monde*. Paris: Mare et Martin, pp. 381–398.

Lormeteau, B. and Molinéro, L. (2018) *Autoconsommation collective et stockage de l'électricité. Etude juridique*, Research report, Paris: ADEME.

Mabile, S. and de Cambiaire, F. (2019) 'L'affirmation d'un devoir de vigilance des entreprises en matière de changement climatique', *EEI*, 5, p. 21.

Marcou, G., Eiller, A.-C., Poupeau, F.-M. and Staropoli, C. (eds) (2015) *Governance and innovations in the energy system. New challenges for local and regional authorities?* Paris: L'Harmattan.

Mathevet, R. (2012) *La solidarité écologique. Ce lien qui nous oblige*. Arles: Actes Sud.

Mathevet, R. and Bousquet, F. (2014) *Résilience et environnement. Penser les changements socio-écologiques*. Paris: Buchet/Castel.

Michelot, A. (2018) 'La solidarité écologique ou l'avenir du droit de l'environnement', in Misonne, D. (ed.) *À quoi sert le droit de l'environnement ? Réalité et spécificité de son apport au droit et à la société*. Brussels: Bruylant, pp. 27–45.

Nadaï, A., Debourdeau, A., Labussière, O., Régnier, Y., Cointe, B. and Dobigny, L. (2015) 'Les territoires face à la transition énergétique, les politiques face à la transition par les territoires?', in Laville, B., Thiébault, S. and Euzen, A. (eds) *Quelles solutions face au changement climatique?* Paris: CNRS éd, pp. 19–36.

Ostrom, E. (2010) *La Gouvernance des biens communs*, 1st edn. Paris: De Boeck.

Petit, Y. (2016) 'Le droit international de l'environnement à la croisée des chemins: globalisation *versus* souveraineté nationale', *RJE*, 1, pp. 31–55.

Poupeau, F.-M. (2007) 'La fabrique d'une solidarité nationale. Etat et élus ruraux dans l'adoption d'une péréquation des tarifs de l'électricité en France', *RFSP*, 57, pp. 599–628.

Poupeau, F.-M. (2013) 'Quand l'État territorialise la politique énergétique. L'expérience des schémas régionaux du climat, de l'air et de l'énergie', *Politiques et Management Public*, 30(4), pp. 443–472.

Puiseux, L. (1978) 'EDF et la politique énergétique', *Après-demain*, 202–203, p. 19.

Riollet, M.Y. and Garabuau-Moussaoui, I. (2015) 'L'énergie fait-elle communauté en France? Le cas de la démarche d'autonomie énergétique du Mené', in Zélem, M.-C. and Beslay, C. (eds) *Sociologie de l'énergie. Gouvernance et pratiques sociales*. Paris: CNRS ed., p. 175.

Rumpala, Y. (2015) 'Formes alternatives de production énergétique et reconfigurations politiques. La sociologie des énergies alternative comme étude des potentialités de réorganisation du collectif', in Zélem, M.-C. and Beslay, C. (eds) *Sociologie de l'énergie. Gouvernance et pratiques sociales*. Paris: CNRS ed., p. 41 et seq.

Sablière, P. (2016) 'La transition énergétique dans les territoires', *JCP A*, n°42, [online].

SNBC (2020) *La transition écologique et solidaire vers la neutralité carbone*, Paris: [online].

Stahl, J.H. (2010) 'L'expérimentation en droit français: une curiosité en mal d'acclimatation', *RJEP*, n 681, Décembre 2010, étude 11.

TEPOS, http://www.territoires-energie-positive.fr/

Terneyre, P. and Boiteau, C. (2017) 'Existe-t-il un droit de l'énergie?', *RFDA*, 3, p. 517.

Van Lang, A. (eds) (2018) *Penser et mettre en oeuvre les transitions écologiques*. Paris: Mare et Martin.

Walker, B., Holling, C.S., Carpenter, S.R. and Kinzig, A. (2004) 'Resilience, adaptability and transformability in social–ecological systems', *Ecology and Society*, 9(2), p. 5.

9 The feasibility of a 100% renewable energy scenario at the village level in Japan from an economic standpoint

Takuo Nakayama

Introduction: the potential for stimulating the local economy through a renewable energy business

Recently, in Japan, much attention has been drawn to stimulating economic viability while enforcing the United Nations' Sustainable Development Goals (SDGs). As the SDGs have been drafted and proposed as a framework for solving global challenges through state-level actions, it needs some adjustment to achieve the goals at the local level.

The Cabinet Office of the Japanese government has tried to apply SDGs at a local level by designating 29 local authorities across Japan as SDG Future Cities and SDG Models in June 2018. In addition to these national projects, local authorities' attempts at their initiatives are increasing. This suggests that sustainable green energy and economic development are increasingly sought at the local authority level. It has been increasingly realized that promoting locally led renewable energy businesses is important in these developments as they contribute to local economic development.

Sustainable Development Goal 7 is "affordable and clean energy", which aims to ensure access to affordable, reliable, sustainable and modern energy for all. Other Goals include: "8. Decent work and economic growth"; "11. Sustainable cities and communities"; "9. Industry, innovation, and infrastructure"; "12. Responsible consumption and production"; and "13. Climate action". All are relevant to local energy development.

Distributed renewable energy that is not dependent on fossil fuels encourages many businesses enterprises to enter the market for the first time, particularly in the electricity generation market, which is supported by the feed-in tariff (FIT) system. This has led to many opportunities for community entities to enter the market. When the system started in 2012, renewable electricity generation gained attention because of the FIT purchase price.

A distributed renewable energy business by a local community or local authority emits less CO_2 as it does not depend on imported fossil fuel or nuclear power; therefore, its carbon load is small in so far as it uses renewable energy. The major concern here is that by entering the electricity generation market, the local community and local authorities are given new opportunities to generate new

DOI: 10.4324/9781003025962-14

wealth and to start economic activities that contribute to sustainable community development.

This chapter shows, using our local value-added analysis model, that small-scale hydroelectric power generation, solar power electricity generation and woody biomass heat generation businesses have generated local economic effects in Nishiawakura village, Okayama prefecture, with a population of about 1,500 in this mountainous area. Furthermore, the chapter constructs a scenario for 100% sourcing of electricity and heat from renewable energy and simulates the resulting standard of local value added.

The economic model to analyze local value added

This chapter is based on an analytical model of local value added to estimate the economic effects of renewable energy at the regional level, particularly at the level of local authorities. The model was largely developed in Germany, which has a rich experience in the spread and introduction of distributed renewable energy, and has been used elsewhere in the world (e.g., Heinbach et al., 2014; Hirschl et al., 2010; Kosfeld and Gückelhorn, 2012; IfaS and DUH, 2013; Hoppenbrock and Albrecht, 2009; BMVBS, 2011).

It is, however, quite a challenging task to measure local economic effects. In Germany, analyses have accumulated on the expansion of renewable energy and the resulting economic effects as a result. However, most of these analyses are conducted at the state levels, and, until recently, there have been very few studies that precisely measure the economic effects at the local authority level.

Studies on the local economic effects generated by renewable energy are gradually emerging in Japan. The majority use the input–output model (Nakayama et al., 2015). They provide analysis based on the interindustry matrix at the state, prefectural or ordinance-designated city level (Konagaya and Maekawa, 2012). Yet, these frameworks lose accuracy when the matrix is adapted for the municipality level.

To solve these problems, some research institutions, including Institut für ökologische Wirtschaftsforschung (IÖW) in Berlin, have developed a model to precisely measure the local economic value added by using the "value chain" advocated by Porter (1985) (Hirschl et al., 2010). This is similar to the gross regional product in production.

Currently, the IÖW model contains many various value chains from a typical portfolio consisting of a distributed power source, heat utilization facilities, transport/supply of biofuel, heat and electricity cogeneration by woody fuel to district remote heat supply (Heinbach et al., 2014). In other words, this can be applied to local authorities in Germany. The model is designed as per German standards and data peculiar to Germany such as the firm's profitability, productivity market, wage level and the German taxation system.

If data particular to another country is available, and if the taxation system can be applied, the approach can be transferred to a different national context. This chapter applies the methodology by using data available in Japan. Consequently, we have applied and modified the model to suit the Japanese context and made it applicable to its local authorities. We call it the local value-added analysis model, and it enables detailed analyses to identify renewable contributions to the local economy (e.g., Morotomi, 2019; Nakayama, 2018; Ogawa and Raupach, 2018; Sando, 2017; Nakayama et al., 2016; Raupach et al., 2015; Raupach-Sumiya et al., 2015; Raupach-Sumiya, 2014).

Considering the various stages in the life cycle of renewable energy facilities, the value chain is generally divided into four phases, namely the "system production phase", which is accounted for only once; the "planning/introduction phase"; the "operation and management phase", which continuously occurs every year; and the "system operator phase". Each of these four phases is further divided into various value chain steps based on the kind of technology.

The current model classifies value added into three elements. They are (a) business profit after tax, (b) employees' disposable income and (c) local tax revenue. The amount of local value added in the renewable energy business for the local authority is the sum (accumulation) of these three elements.

The evaluation of the renewable energy business by the local value-added analysis model

The renewable energy business at the village level

The principal policy for distributed renewable energy business is to use locally available renewable energy sources such as sunlight, wind, hydro-origin, local biomass resources and geothermal energy.

In agricultural communities in mountainous areas surrounded by steep mountains, there is great potential for small-scale hydroelectric power and heat generation using woody biomass (e.g., Nakayama, 2016; Nakayama et al., 2016; Nakayama, 2015). The hydroelectric power generation output depends on the degree of difference between the points of intake and the power generation and the volume of available water flow, where the steep mountainous environment is advantageous in harnessing power. In contrast, if one can obtain woody biomass resources as a byproduct of forestry, they can be utilized as a source of heat.

This chapter conducts an analysis of local value added in Nishiawakura village, Okayama prefecture, which is actively developing renewable energy businesses based on the data provided by the village. Nishiawakura village is well known as an "Environmental Model City" designated by the Ministry of Environment, Japan, which aims to create a low-carbon model area. The village pioneered in introducing a small-scale hydroelectric power generation business and woody biomass boilers as a source of stable income, and it aims to achieve 100% energy self-sufficiency by promoting solar power generation, solar heat utilization and the introduction of electric vehicles.

In this analysis, capital investment occurs only once at the beginning when implementing a renewable energy business as the "planning/introduction phase". The "operation and maintenance" and "system operator" phases are assembled as the "operation/maintenance and business management phase" to include the start of the operation of the renewable energy facility to when it is closed after completing its operational period and expressed as an accumulated value.

In the "planning/introduction phase", we deal with capital investment, which occurs once at the beginning when a renewable energy business is established. In this phase, the planning and designing of the business and facilities, the purchase of facilities and construction work related to the installation of facilities are conducted. In addition, the repayment schedule during the business period is formulated considering the business's capital structure and borrowing conditions. The local value added that occurs in this phase consists of employees' income and business profit from activities that the business in the subject locality conducts such as planning/designing and installation work, and both resident and business taxes are imposed on the income and profit thus generated.

In contrast, the "operation/maintenance and business management phase" refers to the period up to the point when the renewable energy facility completes its operation period and is closed. In this phase, we estimate the annual energy production, sales and maintenance costs individually, and produce the cash flow of each renewable energy business. This will show the value added by the renewable energy business itself and that by adding the business's capital structure (inside–outside locality ratio), we can estimate direct value added by the business to the locality.

Also, we can estimate employees' income and business profit in the locality from the payment to businesses in the locality as subject to analysis such as the payment for woody fuel from maintenance costs of the renewable energy business. Furthermore, based on employees' income and business profit of the renewable energy business and business-related to it in the locality, we can estimate both resident and business taxes. We also individually estimate the annual amount of taxes that occur regardless of whether the business is in profit or not, such as a fixed asset tax.

Electricity generation business (small-scale hydroelectric power generation, solar power generation)

In the village, two types of electricity generation are undertaken: small-scale hydroelectric power generation and solar power generation. A small-scale hydroelectric power plant, M (290 kW), a hydroelectric power plant, K (5 kW), N Ohisama Power Plant (48.6 kW) and Michino-eki solar power plants (20 kW, 20 kW and 15 kW) are operating in the village. A new small-scale hydroelectric power plant, O, is under construction and will begin operations in FY2020.

- Small-scale hydroelectric power plant

The small-scale hydroelectric power plant, M, is a small-scale hydroelectric power plant with an installed capacity of 290 kW. It began operations in 2014 having refurbished the existing plant. The estimated capital investment cost is

300 million yen, and all of this is covered by the village's budget. Expenditure includes subsidies, but these were for the cost to conduct a preliminary examination, and the business itself is fully covered by the village. As the FIT system is applied, generated electricity will be sold at 29 yen/kWh for a period of 20 years. Figure 9.1 presents the analysis results.

The analysis has shown that the business creates local value added three times and more of investment and maintenance cost. Compared with the size of investment and maintenance costs assumed in the FIT system, the actual cost is smaller. This is thought to be essential in creating large value added. Small-scale hydroelectric power plants can continue their operations beyond 20 years, and it is expected that the business will keep bringing value added to the locality for a long time.

As facilities related to hydroelectric power generation contain some that have incurred more than 20 years of depreciation is at 38 years or 57 years in the figure provided, the sizes of the investment and depreciation do not match. Essentially, it can be assumed that the plan will continue generating electricity on a stable basis after 20 years of its operations, and it can be expected that it will continue producing local economic value added by selling electricity. However, in the 21st year of operations, the FIT system's application will cease. This means selling electricity, including that for internal consumption, will become a challenge to be tackled.

Moreover, there is a K hydroelectric power plant with an installed capacity of 5 kW in the village. This has been installed for disaster prevention and began operating in 2016. The amount required for capital investment is said to be 33 million yen, which was entirely covered by the village. The FIT system is applied, and generated electricity is sold at 34 yen/kWh for 20 years.

When we analyze this facility, it only generates local value added to cover about 40% of the investment and maintenance. This is because the business is not profitable as the sales estimated for the period of 20 years are smaller than the amount of investment. However, as the facility is built for disaster prevention and

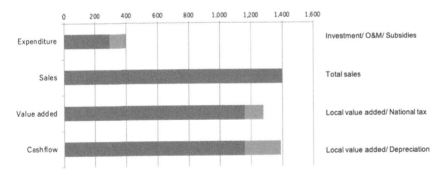

※ Accumulation from the installation to the 20th year of operation

Figure 9.1 The amount of local value added by a small-scale hydroelectric power plant, M (unit: million yen).

not for achieving economic effects on its own as a renewable energy business, whether it is appropriate or not needs to be discussed elsewhere.

Furthermore, a specific-purpose company to be established afresh is planning to operate a small-scale electricity generation plant with an installed capacity of 199 kW in 2020. The size of the capital investment necessary is estimated to be about 450 million yen. The plan is that all the necessary investments will be made by the firms in the village, and two-thirds of necessary funds will be procured by borrowing from financial institutions to implement the project. The FIT system is to be applied, and generated electricity will be sold at 34 yen/kWh for a period of 20 years.

The analysis has shown that the business will generate about 300 million yen of local value added accumulated from the installation to the 20th year of operation. The cost of dismantling reserves paid annually is relatively large, at 4% of capital investment, which stretches business profit. In contrast, the personnel cost for maintenance and management is small and as the income of employees of the business and related businesses is small, the source of local value added at the operation/maintenance and business management phase is mainly the business's (the electricity generation company's) net profit and local tax such as fixed asset tax.

As the size of electricity generation, how the business is managed, how it is financed and the price at which electricity is sold differ, each electricity generation plant has a different cost structure. Figure 9.2 totals the expenditure, sales and value added by the three small-scale electricity generation plants.

Even when we combine small-scale hydroelectric power plant K with poor profitability and small-scale hydroelectric power plant O with high operation and maintenance costs relative to M small-scale hydroelectric power plant, the three power plants in the village achieve sales about 1.6 times the cost required for investment and operation/maintenance. Local value added realized by these

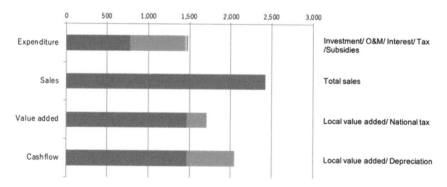

※ Accumulation from the installation to the 20th year of operation

Figure 9.2 The amount of local value added by three small-scale hydroelectric power plants in the village (unit: million yen).

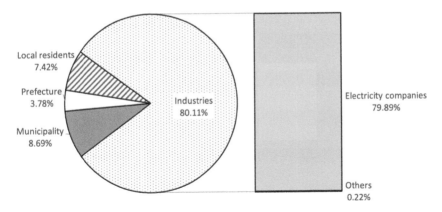

※ Accumulated value till the 20th year of the operation

Figure 9.3 Attribution of local value added during the operation and maintenance (O&M) phase of the three power generation businesses in the village (after income tax).

businesses in the village is estimated to be about 1.5 billion yen from the installation until the 20th year of operation.

Figure 9.3 shows the attribution of local value added by the three small-scale hydroelectric power generation businesses in the village. The figure shows, in percentage, the attribution of value added considering the business profit and local taxation revenue after employees' income. About 80% of it is attributed to the electricity companies (power generation businesses) in the industry sector. Others in the industry sector include businesses that conduct repair and maintenance. The tax revenue of the local municipality and prefecture is more than 14%, and more than 7% is attributed to residents who invested in the business (Figure 9.3).

- **Solar power generation business**

The N Ohisama Power Plant is a solar power generation plant with an installed capacity of 48.64 kW. It is installed on the roof of the Nishiawakura Convention Hall and has been operating since 2014. Nishiawakura village rents out the roof free of charge, and it is used as an emergency power source at times of disaster and for environmental education.

The cost of installing the facility has been met by 49 million yen from 28 residents and a loan of 100 million yen from a regional Okayama prefecture bank. The facility is run by a non-profit organization (NPO) of Okayama city and part of the revenue is used to support projects to vitalize the mountainous area. The FIT system is applied to this business as well and generated electricity is sold at 36 yen/kWh for a period of 20 years. Figure 9.4 shows the analysis of this business.

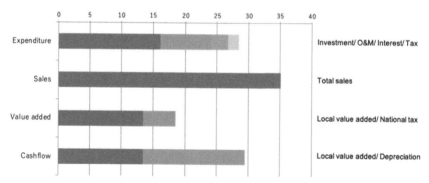

※ Accumulated value from the installation till the 20th year of operation

Figure 9.4 The amount of local value added by the N Ohisama power generation plant (unit: million yen).

The analysis has shown that the business creates local value added, which accounts for more than 50%of the capital investment and operation/maintenance costs. The sales are also estimated to be bigger than total investment expenditure, operation/maintenance cost, interest and tax, which means that the business is profitable. As for the distribution of value added, the business's net profit is the largest, and as this is attributed to the investors in the village who own the business, the business is valuable for the local community.

The annual operation/maintenance cost is about 1.78 million yen, of which 1.48 million yen is paid outside the local community as "the difference between other costs and income". This is presumably a payment to the NPO that runs the business.

In another solar power generation business, the Michino-eki, as in the case of N Ohisama Power Plant, the village rents out the roofs, and there are two facilities of 20 kW and one of 15 kW. Funds have been procured again like in N Ohisama Power Plant from local residents and a loan from a financial institution.

Figure 9.5 shows the total value added by the solar power generation plants in the village where the village rents out the roof. It shows that the business is generally profitable. In contrast, it is estimated that 18 million yen of local value added will be accumulated in the village in the period from the installation to the 20th year of operation.

Figure 9.6 shows the attribution of the local value added of the operation/maintenance phase of the solar power generation business after employee salaries. As the business is a so-called rent-a-roof business, the proportion attributed to the industry sector is lower than that of the other power generation businesses, at 36%. In contrast, the proportion of donation/tax revenue attributed to the municipality/prefecture is about 45%. The return on residents' investment accounts for about 20%, which is higher than that of small-scale hydroelectric power generation.

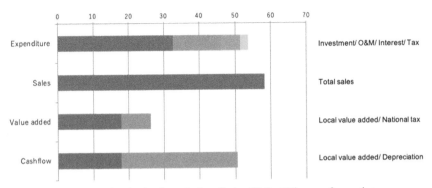

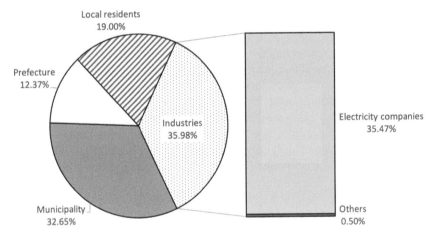

※ Accumulated value from the installation till the 20th year of operation

Figure 9.5 The total local value added of solar power generation business (unit: million yen).

※ Accumulated value from the installation till the 20th year of operation

Figure 9.6 Attribution of local value added during the operation phase of the solar power generation business (after employees' income).

Biomass heat supply business

Nishiawakura village is engaged in a heat supply business fully utilizing woody biomass. This section first conducts an analysis of the local value added by the wood-fired boiler installed in the public bath facility Y in the village to exclusively supply heat (340 kW). It was installed and began operating in 2015.

The business' capital structure is estimated to consist of investment from local businesses accounting for 30% and a loan from a financial institution outside the village accounting for 70%. Village data provide that the loan period is 12 years, and the borrowing interest rate is at 0.1%, which is a very small interest payment.

The boiler is owned by Nishiawakura village, which also owns the facility. The facility itself is run by Company A (the heat consumer), the designated administrator, and the boiler is run by Company S (the heat supplier). Company A pays for maintenance of the boiler and the electricity for the facility; Company S, the heat supplier, pays for the other cost relevant to the operation of the boiler such as biomass fuel cost.

The unit price of sales of heat from the heat supplier S to the heat consumer A (since June 2018) is at 2,220 yen/GJ, and when maintenance cost and electricity bills, which the heat customer A pays are added, the unit price of purchase of heat for the heat customer A becomes 2,625 yen/GJ.

In this analysis, we included maintenance cost and electricity bills, which the heat consumer A pays as cost of the heat supply business to evaluate the profitability and local value added of the heat supply business; simultaneously, we used 2,625 yen/GJ, the actual unit purchase price for the heat consumer A, as the unit sales of heat price. Considering the cost incurred in operating and maintaining the facility, all biomass fuel (wood) is supposed to be purchased from the village.

Figure 9.7 shows the analysis results. The analysis shows that the business produces local value added, which is more than 70% of the total investment, as costs are necessary for operation/maintenance, tax and subsidies. The accumulative sales exceed expenditure, which shows that the business is profitable.

Figure 9.8 shows the chronological change in the accumulative value of local value added and its constituent components. The business received a 40-million-yen subsidy, and it shows that by 2020 – the sixth year since the start of operation – the business produced more local value added than what it received as subsidies. The largest proportion of local value added is the profit generated by the business. However, the cost of biomass fuel, part of the operation/maintenance cost, is paid

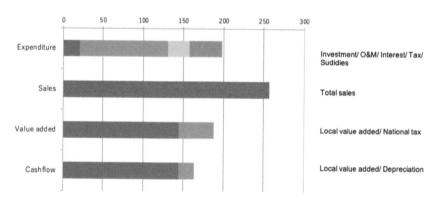

※ Accumulated value from the installation till the 20th year of operation

Figure 9.7 The amount of local value added by the wood-fired boiler (340 kW) (unit: million yen).

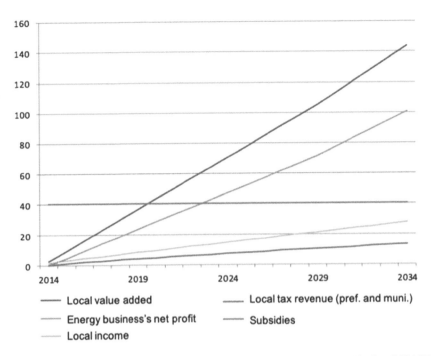

Figure 9.8 Changes in accumulative local value added by the wood-fired boiler (340 kW) (unit: million yen).

to local agriculture/forestry businesses, which then generates value added in these areas and a degree of local income.

In the village, there are accommodations K with wood-fired boilers of 270 kW (170 kW + 100 kW, which began operating in 2016), and accommodation M with a wood-fired boiler of 75 kW (which began operating in 2015). These heat supply businesses are operated according to the same scheme as the public bath facility Y with a wood-fired boiler (340 kW). Figure 9.9 shows the total local value added by the three woody biomass heat supply businesses (the public bath facility Y, accommodation K and accommodation M) in the village. Figure 9.10 shows the distribution of local value added during the operation/maintenance phase.

As Figure 9.9 shows, the analysis has found that the three businesses will accumulate about 280 million yen of local value added by the 20th year of operation. At the investment phase, equipment (the boiler), which accounts for the majority of the cost, is purchased from outside the village and the amount of local value added due to designing and construction work is small compared with that of the operation/maintenance phase. In contrast, in the operation/maintenance phase, local value added is created by business profit, and the agriculture/forestry sector, which supplies fuel, is to have an accumulative value that will reach about 270 million yen.

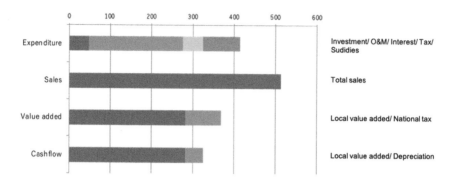

Figure 9.9 The total local value added by the three businesses in the village (unit: million yen). ※Accumulated value from the installation until the 20th year of operation. Source: Based on data from Nishiawakura village.

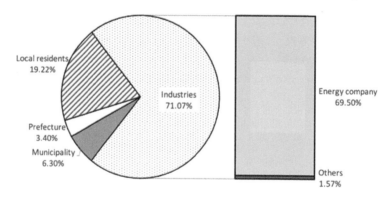

※ The total of local value added during the operation/maintenance phase: about 270 million yen

Figure 9.10 Attribution of local value added by the three businesses in the village (after employees' income). ※The total local value added during the operation/ maintenance phase: about 270 million yen. Source: Based on data from Nishiawakura village.

Figure 9.10 shows that the value added attributed to the heat supplier (Company S), which runs the heat supply business, is largest at 70%. This is followed by residents, and they are engaged in forestry that supplies wood as fuel.

These show that in Nishiawakura village, the biomass energy business is profitable, and it is expected that the business will further expand using generated value added (Figure 9.10).

The scenario to realize a 100% renewable energy local authority

Nishiawakura village drafted the "New Energy Vision for the Community" in 2005 and set its goal to achieve 100% energy self-sufficiency. The village has since been actively working to achieve 100% energy self-sufficiency by renewable energy through Environmental Model Cities and Biomass Town projects. As we have been reviewing, several projects related to power generation and heat supply at a reasonable scale are being implemented in Nishiawakura village.

As for power generation, a plan to start a new small-scale hydroelectric power plant O of 199 kW in 2021 is being pursued. According to the town hall, this power plant can meet 70% of the village's electricity demand. As for community heat supply, a community heat supply system that generates heat by a woody biomass boiler (about 400 kW) and supplies it in the form of heating/hot water in the center of the village, including the town hall, is being pursued. According to the town hall, when the community heat supply system is complete, nearly 40% of the community's heat demand will be met.

As seen earlier, Nishiawakura village has experience in implementing projects in power generation and heat supply businesses. Consequently, it is appropriate to construct a business-as-usual (BAU) scenario based on experience. It is essential to estimate the village's further demand to construct a scenario for 100% self-sufficiency. First, we examine the household sector.

According to the Population Decline Committee of the Japan Policy Council's estimate, 14 municipalities in Okayama prefecture were facing the risk of extinction and Nishiawakura village was also classified as facing possible extinction. This article shocked Japan when it was published in 2014.

However, according to statistics based on the basic resident register, while the population declines slightly every year, the number of households is slightly increasing. We can assume a societal increase factor behind it. As far as the tendency over the recent six years, we cannot predict a huge decrease in the population/number of households in the future, and, consequently, we predict little decline in energy consumption such as electricity and heat in the household sector.

In contrast, in the industry sector, the growth of new local venture wood processing businesses using locally produced woods have been reported in many places. As these businesses require a certain amount of energy, we do not think energy demand in the industry sector will fall significantly. Following these, the analysis in this section does not consider a further decline in the energy demand in the village.

One hundred percent sourcing of electricity from renewable energy

As for electricity, consider that 100% sourcing from renewable energy is possible while taking the annual amount of power generation (Wh) as the unit. As highlighted previously, Nishiawakura village has two small-scale hydroelectric power plants in operation producing 295 kW (290 kW from the small-scale hydroelectric power plant M and 5 kW from the small-scale hydroelectric power plant K). With a small-scale hydroelectric power plant O with 199 kW, the village will have a capacity of 494 kW.

The capacity factor of the small-scale hydroelectric power plants in Nishiawakura village is high at around 86–90%. However, in a few years, power generation can decrease due to drought and so on; therefore, the capacity factor of the small-scale hydroelectric power plants in the village is set to 80%, which translates to about 3.5 GWh ($\fallingdotseq$ 494 kW × 365 days × 24 hours × 0.8). Essentially, for each kW generated by the small-scale hydroelectric power plants, about 7 MWh ($\fallingdotseq$ 3.5 GWh/494 kW) of electricity is generated annually.

Furthermore, the village is already capable of producing 104 kW ($\fallingdotseq$ 48.64 kW + 20 kW + 20 kW + 15 kW) through solar power generation facilities. This means that about 100 MWh (i.e., 0.1 GWh) of electricity is generated annually. According to the town hall, the N Ohisama Power Plant with a capacity of 18.64 kW can generate 59 MWh of electricity annually. Basically, for each kW of the solar power generation plant's capacity, about 1 MWh ($\fallingdotseq$ 50 MWh/48.64 kW) of electricity is generated annually.

Essentially, under the sunlight conditions of Nishiawakura village, if the village were to install facilities such as a panel power conditioner for solar power generation with the same capacity as the N Ohisama Power Plant, they need solar power generation facilities of about 1 kW to generate 1 MWh of electricity annually.

Combined, the small-scale hydroelectric power plants and solar power plants currently produce 3.6 GWh (3.5 GWh + 0.1 GWh) of electricity annually. According to the town hall, when the small-scale hydroelectric power plant O begins operations, the village will achieve 70% self-sufficiency in electricity. The village can thus generate 5.1 GWh ($\fallingdotseq$ 3.6 GWh/0.7) of electricity annually, and it can achieve 100% self-sufficiency in electricity.

Furthermore, we would like to propose a feasible scenario to cover the current gap of 1.5 GWh (= 5.1 GWh − 3.6 GWh) to achieve 100% sourcing from renewable energy by solar power generation.

To generate 1.5 GWh more electricity annually to achieve 100% sourcing of electricity from renewable energy, it is necessary to install solar power generation facilities with the installed capacity of 1.5 MW at the same standards as the N Ohisama Power Plant.

One hundred percent sourcing of heat supply from renewable energy

Currently, the village has woody biomass heat supply facilities (wood-fired boilers) with an installed capacity of 685 kW (340 kW + 170 kW + 100 kW + 75 kW). According to Nishiawakura village, once a woody biomass boiler for the community heat supply (400 kW), due to begin operations in FY2020, is introduced, the combined capacity of about 1.1 MW ($\fallingdotseq$ 685 kW + 400 kW) can cover 40% of the village's head demand.

To meet the village's demand for heat by renewable energy, they need a total capacity of about 2.7 MW ($\fallingdotseq$ 1.1 MW/0.4). It then follows that what is needed is a new biomass heat supply system with a capacity of 1.6 MW (= 2.7 MW − 1.1 MW).

So far, our calculations have shown the installed capacity of the power plants and heat supply facilities necessary to achieve 100% sourcing from renewable energy. For the solar power generation plant, facilities must generate 1.5 GWh of electricity per year, and if this is to be met by solar power generation, an additional installed capacity of about 1.5 MW is needed at the same technical standards as the existing N Ohisama Power Plant. As for the heat supply facilities, an additional output of 1.6 MW at the same standards as the current ones is needed.

The simulation of local value added for a 100% renewable energy local authority

The amount of local value added by the introduction of 100% sourcing from renewable energy (electricity)

The analysis in the previous section suggests that the three small-scale hydroelectric power plants (small-scale hydroelectric power plant M, small-scale hydroelectric power plant K and small-scale hydroelectric power plant O) provide the village with a total installed capacity of 494 kW, which generates an accumulative total of 1,463 million yen over 20 years. In contrast, considering solar power generation, the two projects (N Ohisama Power Plant and *Michino-eki* solar power plants) provide the village with an installed capacity of 104 kW and generate an accumulative total of 18 million yen of local value added over 20 years. Combined, these two generate 1,481 million yen ($\textcircled{1}$) over 20 years.

   Suppose that achieving 100% sourcing of electricity from renewable energy in Nishiawakura village requires a new solar power generation facility with a capacity of 1.5 MW to be built. Solar power generation has been spreading greatly in recent years and, with it, the coordinated price offered in the FIT system has decreased considerably. Here it is assumed that the introductory cost of the facility has also gone down sufficiently and that the new facility would generate local value added at the same level of profitability of N Ohisama Power Plant, which is the existing facility.

   The analysis in the previous section has shown that N Ohisama Power Plant generates an accumulative total of 260,000 yen/kW ($\fallingdotseq$ 13 million yen/48.64 kW) of local value added over the 20 years. If the new facility is to be introduced at the same standard, it is estimated that with solar power generation of 1.5 MW, it could generate an accumulative total of 39 million yen ($\fallingdotseq$ 1,500 kW × 260,000 yen/kW) ($\textcircled{2}$) of local value added.

   Therefore, an accumulative total of 1,871 million yen ($\textcircled{1}$ + $\textcircled{2}$) of local value added over the period of 20 years is estimated in the power generation sector.

The amount of local value added by the introduction of 100% sourcing from renewable energy (heat)

Currently, three wood-fired boilers to supply heat (public bath facility Y, accommodation K and accommodation M) provide the village with a total installed capacity of 685 kW. When the woodchip boiler for community heat supply (400 kW), currently under construction, is introduced, the installed capacity of heat supply facilities would be about 1.1 MW, which is about 40% of the village's heat demand.

The analysis in the previous section has shown that the size of local value added by the wood-fired boilers (685 kW) currently in operation, would be an accumulative value of 282 million yen over 20 years. This gives us a value of 411,700 yen/kW ($\fallingdotseq$ 282 million yen/685 kW) (accumulative over 20 years).

If we suppose that the woodchip boiler/community heat supply project (400 kW) currently under construction is run by the same heat supplier that operates the existing wood-fired boilers and that value added will be generated at the same level, an accumulative total of 165 million yen ($\fallingdotseq$ 417,700 yen/kW × 400 kW) of local value added is estimated to be generated over 20 years.

The local value added generated by the total of about 1.1 MW consisting of the wood-fired boilers (685 kW) currently in operation and the chip boiler/community heat supply (400 kW) would be 447 million yen (③) over the 20-year period.

If we assume that the capacity of a woody biomass boiler to be introduced in the future to realize 100% sourcing from renewable energy is 1.6 MW, the amount of local value added under the same conditions as the existing wood-fired boilers would be 659 million yen ($\fallingdotseq$ 411,700 yen/kW × 1,600 kW) over the period of 20 years (④). When these are added, the accumulative value of local value added by realizing 100% sourcing from renewable energy in the heat sector would be 110.6 billion yen (③ + ④) over the 20-year period.

Section summary

The current section has evaluated the installed capacity of electricity and heat sectors at the village level necessary to achieve 100% sourcing of electricity and heat from renewable energy and has simulated the amount of local value added when sourcing is achieved.

As for the electricity sector, we have estimated how much local value added would be gained if 100% sourcing from renewable energy is achieved by adding a new solar power plant of 1.5 MW to the existing/planned power plants. The result shows that an accumulative value of 1,871 million yen (⑤) over 20 years would be generated, including by existing power plants.

In the heat sector, we have estimated how much local value added would be achieved if 100% sourcing from renewable energy is acquired by adding a new woody biomass heat supply facility of 1.6 MW to the existing/planned facilities. The result shows that an accumulative value of 1,106 million yen (⑥) over 20 years would be generated, including by existing wood-fired boilers.

Adding these together, 2,977 million yen (⑤ + ⑥) of local value added is to be generated over 20 years. When simply divided by 20 years, we obtain a value of 148.85 million yen/year (= 2,977 million yen/20 years).

Summary

Using local value-added analysis, which can precisely measure the local economic effects of renewable energy at the local authority level, we have estimated the amount of local value added of the small-scale hydroelectric power generation business and woody biomass heat supply business in Nishiawakura village, Okayama prefecture. In the small-scale hydroelectric power generation business, owing to the small-scale hydroelectric power plant M as a replacement project, the economic effects are as estimated or more.

Notably, in this chapter, the woody biomass heat supply business is feasible in the local economic setting. The initial cost is certainly incurred when switching from the fossil fuel boiler system to the wood-fired one. As these boiler systems are not necessarily produced locally in a small village, introducing technology based on advanced knowledge from outside the village (or abroad) is rational. Therefore, it is inevitable that some subsidies are paid. In contrast, we should focus on the fact that after about five years of operation, the accumulative value of local value added exceeds that of subsidies. This shows that the subsidies have worked as a stimulant to revitalize the regions in the context of the sustainable development of local communities. In the subsequent operation/maintenance phase, local value added by the project continues to be accumulated annually throughout the facility's service life. If some of the local value added is set aside as a fund, it could be used to renew the facility next time.

Besides, in Nishiawakura village, the local government took the lead in vertically unifying the value chains of related stakeholders from the procurement of wood, wood processing, wood drying, wood supply and wood-fired boiler management so that there is a merit of scale at the appropriate size. The local business model wherein heat is sold to the final heat consumer by the thermal unit (J) realizes a local cyclical economic model at the initiative of the locality in addition to making use of the support of the FIT system (electricity) by the central government. We should also focus on the fact that local actors were consulted numerous times to ensure that local characteristics acquired economic rationality. Various talents, including representative Dr. I, who has provided ideas and led the implementation, have played an important role in Company S (an energy (heat supply) business).

The current chapter has empirically shown that even a small village population of about 1,500 in a mountainous area can build a sustainable developmental model for the community that sufficiently contributes to the realization of SDGs on the environment, energy and community development by adapting renewable energy businesses to local standards. Furthermore, the scenario of 100% sourcing from renewable energy has estimated that a total of 2,977 million yen of local value added would be accumulated over 20 years. For one year, it is estimated to be 148.85 million yen per year. As the amount of local value added is a total of employees' disposable income, business profit after tax and local tax revenue, we cannot conduct a simple comparison. However, to understand this figure, if we look at Nishiawakura village's annual revenue for FY2018, of its self-sourced revenue, 137.4 million yen is municipality tax and 386.45 million yen is from other self-sourced funds.

As for the electricity sector, the chapter has shown that 100% sourcing of electricity from renewable sources can be achieved. However, the system price of solar power generation has recently decreased dramatically. In this analysis, the amount of value added estimated from the experience of running the business since 2015 is used in building the BAU scenario, and it is also appropriate to consider the possibility of on-site power generation for local consumption, which reflects the latest situation.

In contrast, achieving 100% sourcing of heat supply from renewable energy would require more than double the existing installed capacity. We need to have a detailed study if enough biomass resources are to be procured within the village to achieve this. Moreover, most businesses in the village with substantial demand will be covered by the projects that are already introduced or to be introduced. We must consider the possibility of heat supply by electricity to meet the demand for heat that is further diffused.

The construction of the BAU scenario to realize 100% sourcing from renewable energy and the analysis of local value added in the current chapter have shown a degree of feasibility and adherence to the standards of local value-added creation. However, the analyses in the chapter have only shown the grounds on which the actual scenario planning takes place.

In executing renewable energy projects, the local authority and the community itself must gain more income in the future with a sustainable scenario and economic effects. We hope that future researchers develop a more concrete and comprehensive scenario to realize 100% sourcing from renewable energy that is unique to the community and is supported by residents.

References

BMVBS (Bundesministerium für Verkehr, Bau und Stadtentwicklung) (2011) 'Strategische Einbindung regenerativer Energien in regionale Energiekonzepte - Wertschöpfung auf regionaler Ebene', https://www.bbsr.bund.de/BBSR/DE/Veroeffentlichungen/BMVBS /Online/2011/DL_ON182011.pdf?__blob=publicationFile&v=2

Heinbach, K., Aretz, A., Hirschl, B., Prahl, A. and Salecki, S. (2014) 'Renewable energies and their impact on local value-added and employment', *Energy, Sustainability and Society*, 4(1), pp. 1–10.

Hirschl, B., Aretz, A., Prahl, A., Böther, T., Heinbach, K., Pick, D. and Funcke, S. (2010) *Kommunale Wertschöpfung durch Erneuerbare Energien*, Schriftenreihe des IÖW 196/10. Berlin: Institut für Ökologische Wirtschaftsforschung.

Hoppenbrook, C. and Albrecht, A.K. (2009) 'Diskussionspapier zur Erfassung der regionaler Wertschöpfung in 100%-EE-Regionen', *DEENET (Hrsg.)*, *Arbeitsmaterialien 100EE*, No. 2, http://www.100-ee.de/downloads/schriftenreihe/ ?eID=dam_frontend_push&docID=1140

IfaS (Institut für angewandtes Stoffmanagement), DUH (Deutsche Umwelthilfe e.V.) (2013) *Kommunale Investitionen in Erneuerbare Energien – Wirkungen und Perspektiven.* http://www.stoffstrom.org/fileadmin/userdaten/dokumente/Veroeffentlichungen/2013 -04-04_Endbericht.pdf

Konagaya, K. and Maekawa, T. (2012) *Introduction to economic effect: The tool for regional revitalisation, policy drafting and policy evaluation.* Nihon-hyoronsha: Tokyo.

Kosfeld, R. and Gückelhorn, F. (2012) 'Ökonomische Effekte erneuerbarer Energien auf regionaler Ebene', *Raumforsch Raumordn*, 70, pp. 437–449. https://link.springer.com/article/10.1007%2Fs13147-012-0167-x

Morotomi, T. (2013) 'To revitalise the community with renewable energy: What we can learn from Germany shifting to the "distributed electricity system"', *Sekai* (Iwanami-shoten), 2014(10), pp. 153–162.

Morotomi, T. (2019) *Introduction to local value-added analysis: Local circulatory economy promoted by renewable energy.* Hihon-hyoronsha: Tokyo.

Nakamura, R., Nakazawa, J. and Matsumoto, A. (2010) 'CO_2 reduction using woody biomass and its local economic effects: The construction of local input-output model and its new application', *Studies in Regional Science*, 42(4), pp. 799–817.

Nakayama, T. (2015) 'A study of regional economy: Sustainable resources hidden in a marginal settlement', in Morotomi, T. (ed.) *22 jobs that change energy.* Gakugei-shuppansha: Kyoto, pp. 185–192.

Nakayama, T. (2016) 'Can we claw back 1 per cent of income to the mountainous area with renewable energy?', *Public Finance and Public Policy*, 60, pp. 3–17.

Nakayama, T. (2018) 'The economic effects of renewable energy to rural areas', *Kagaku*, Iwanami-shoten, 88(10), pp. 997–1004.

Nakayama, T., Raupach-Sumiya, J. and Morotomi, T. (2016) 'The analysis of local value-added by distributed renewable energy', *Research on Environmental Disruption*, 45(4), pp. 20–26.

Nakayama, T., Raupach-Sumiya, J. and Morotomi, T. (2016) 'The generation of local value-added of renewable energy in Japan: The introduction, test and application of the Japanese version of local value-added analysis model', *Sustainability Research*, 6, pp. 101–115.

Nakayama, T., et al. (B style PJ research group) (2016) 'An economic effect analysis of a small-scale system starting with wood: The construction of community-led system', in *Producing and consuming energy locally using woody biomass heat.* Zenrinkyo: Tokyo, pp. 118–135.

Nakayama, T., Raupach-Sumiya, J. and Morotomi, T. (2015) 'The application of the analysis of local value-added by distributed renewable energy to Japan', Discussion Paper Series, No. 15-B-1, Kyoto University, the 'Research into system design of distributed electricity system, evaluation of its social/economic effects and its contribution to regional revitalisation project'.

Ogawa, Y. and Raupach-Sumiya, J. (2018) 'Economic effects of renewable energy to the community: A case study using value chain analysis', *Environmental Science*, 31(1), pp. 34–42.

Porter, M.E. (1985) *Competitive Advantage: Creating and Sustaining Superior Performance.* The Free Press: NY.

Raupach-Sumiya, J. (2014) 'Measuring regional economic value-added of renewable energy - The case of Germany', *Shakai Shisutemu Kenkyu (Social System Study)*, 29. Ritsumeikan University BKC Research Organization of Social Sciences Kyoto pp 1–31; http://www.rit-sumei.ac.jp/acd/re/ssrc/result/memoirs/kiyou29/29-01.pdf

Raupach-Sumiya, J., Matsubara, H., Prahl, A., Aretz, A. and Salecki, S. (2015) 'Regional economic effect renewable energies – comparing German and Japan', *Energy, Sustainability and Society*, 5(10), a Springer Open Journal. https://doi.org/10.1186/s13705-015-0036-x

Raupach-Sumiya, J., Nakayama, T. and Morotomi, T. (2015) 'Economic effects renewable energy will bring to communities in Japan: A model to calculate using input-output analysis by the power source', in Morotomi, T. (ed.) *Renewable energy and regional revitalisation*. Nihon-hyoronsha: Tokyo, pp. 125–146.

Sando, A. (2017) 'The analysis of local economic value-added in thermal power generation', *Public Finance and Public Policy*, 39(2), pp. 121–130.

10 Actors, motives and social implications of 100% renewable energy territories in Austria and Germany

Laure Dobigny

The emergency of climate change and energy issues (fossil energy depletion and their CO_2 emissions) are driving more and more territories toward implementing renewable energy (RE) on a local scale. Austria and Germany were pioneer countries in developing and implementing RE, on national and local scales, with the result that some local towns in both countries became among the first to achieve complete autonomy with RE in the late 1990s and early 2000s. Named 100% renewable energy territories, they produce more RE than they consume. How do we understand this turn? What are the circumstances, actors and motives of this local RE transition? And what are the social implications of local energy (re) appropriation on the economy, solidarity and identity? Based on a socio-anthropological study of four towns in Austria and Germany, this chapter proposes to provide answers to these questions. To begin with, we will see that local energy transition is led by unconventional players in the energy sector: individuals, farmers and local communities.

In contrast to France and Japan, both countries with a centralized political and energy system, Austria and Germany are both federal countries. Nevertheless, if the federal system plays a role in this local RE implementation, it is less due to a decentralized energy system than to citizen empowerment. Evrard (2013) demonstrates that in Germany, territories have in fact few powers in terms of electricity development and orientation, mostly determined by large energy companies named the 'Big Four' (EnBW, E.ON, RWE, Vattenfall). Nevertheless, due to the federal political system, citizens are used to being involved in local issues. And as we will see in this chapter, local energy autonomy is part of an empowerment process. Thus, if a decentralized energy system does not play a substantial role in local energy transition, the federal political system influences it by empowering citizens. That is one explaining factor of most citizen- and community-based RE initiatives in Austria and Germany compared with France and Japan.

In this way, local RE transition is led by social, political and ecological values. Supporting local stakeholders is a strong motive of RE implementation, as well as the ecological, social and democratic dimensions of RE projects. Indeed, local RE transition has economic and social benefits, such as job creation, the development of energy tourism, new solidarities within and outside communities, symbolic

DOI: 10.4324/9781003025962-15

dimensions like inhabitant recognition and identification, and finally, the emergence of a collective identity around local RE transition. This chapter aims to present and discuss these results.

Methods and case studies

A socio-anthropological approach framed the field study, based on observations and in-depth semi-structured interviews in the native tongue of interviewees. Qualitative methods are the most suitable when it comes to exploring motives, local processes or individual and collective representations. The case selection criteria were that local actors (municipality and/or citizens) led the renewable energy plants and achieved full self-sufficiency for heat and electricity (100% RE territories). Thus, the field study is based on four rural towns from 800 to 6,300 inhabitants in Germany and Austria: Freiamt (Germany, Baden-Württemberg, 4,200 inhabitants), Jühnde (Germany, Lower Saxony, 780 inhabitants), Mureck (Austria, Styria 1,600 inhabitants) and Güssing (Austria, Burgenland, 3,700 inhabitants). The criterion of RE self-sufficiency for heat and electricity explains the small size of towns studied. It is obviously easier and quicker to achieve complete energy self-sufficiency in small towns rather than large towns with more inhabitants, companies, businesses and sometimes industries with higher energy consumption. The fact that these towns are all based in rural areas also reveals the greatest ease for rural towns compared with urban towns to produce RE, which requires sources like biomass, wood, space for wind turbines implementation, rivers, cultivation of oil plants like rape and so on. Energy self-sufficiency was achieved by a mix of RE, depending on local potentialities, but all towns studied implemented biogas plants and biomass district heating (BDH). The biomass source used in these BDH is biogas and/or wood. According to local potentialities, the other RE implemented could be solar thermal panels and/or photovoltaic panels, wind turbines, micro-hydro plants and small biodiesel plants (rape oil). In the four towns, electricity power plants (like wind turbines or biogas plants) are connected to the national grid, and BDH constitutes a local and independent grid. With RE projects launched in the 1990s and 2000s, these communities are pioneer towns of RE transition. The four towns achieved RE autonomy in 2005, sometimes initiated 20 years ago (Mureck) or just four years ago (Jühnde). The difference of temporalities between these towns is due to different factors: (1) social and geographical morphologies play a huge role, as mentioned before, it is easier and quicker for a small and centralized town to achieve energy autonomy; (2) it takes longer for the pioneer towns to bring the relevant techniques into existence (like building an oil mill for around 500 farmers in Mureck) than the next one who could reuse inventions or organizational processes; (3) the influence of regulatory and legislative frameworks differs between countries but also in time, with new legal measures in favor of citizen RE projects, for example; and (4) the role played by energy policies and political structures (Dobigny, 2019).

Circumstances and actors of 100% RE territories

RE transition differs between the four territories, but they all face local difficulties. These difficulties are sometimes cumulative: few economic opportunities or tourism, unemployment and few opportunities of employment, poverty, aging and/or decrease of population, poor soils, pollution and so on. Thus, contrary to what one might think, it is not in wealthy territories that the most innovative RE transitions emerge, but often in territories with difficulties. These difficulties trigger change and the development of new opportunities, which have simultaneously social, economic, ecological and symbolic dimensions. Local difficulties are thus the driving forces of action: they open possibilities of creation and innovation through the mobilization of new ideas and actors. According to a mayor, in a territory without difficulties, one does not think at all of these things.

Circumstances of emergence nevertheless differ between the four towns. In Güssing (Burgenland, Austria), developing RE projects was a political will, led by the municipality, especially the mayor, to avoid poverty, unemployment and the decrease of population. In our panel, surprisingly, that was the only town where RE transition was led by local policymakers. In other towns, RE transition was initiated and led by inhabitants and local stakeholders, especially farmers.

In Mureck (Styria, Austria), local RE transition and autonomy are due to a farmers' initiative. In 1985, three farmers decided to develop a biodiesel plant (rape oil production) and enrolled 500 local farmers in a cooperative to do it (1991). Then, the farmers' cooperative established a local BDH based on wood usage in the town, then a biogas plant producing electricity and heat (connecting to the BDH), and finally a citizen photovoltaic farm.

In Freiamt (Baden-Württenberg, Germany), circumstances are different. Wind farm developers came with a wind turbine implementation project that residents declined. They wanted to undertake the project through a citizen company. After that first project, a lot of small RE plants have emerged, either collectively (e.g., the municipality, associations, sports clubs), by local stakeholders and businesses (e.g., farmers, sawmill) or individually.

In Jühnde (Lower Saxony, Germany), the idea came from a research-action project led by an interdisciplinary team from the University of Göttingen (the Interdisciplinary Center for Sustainable Development, IZNE). They proposed and assisted the inhabitants to reach local energy autonomy by the implementation of a biogas plant. Inhabitants have formed a cooperative to manage the power plant that was partly financed by inhabitants' financial participation, subventions and a bank loan.

Despite these specific and locally based stories, they illustrate most of the cases of local RE transition circumstances: a farmers' initiative, the venue of external RE developers or a municipal initiative. But contrary to our hypothesis, the role of the municipality as leader of local RE transition is low. In our case study, it is only true for one in four towns. In all towns, municipalities have nevertheless supported and promoted local RE projects, unanimously among the different political parties represented. That has had a decisive impact on the realization of such RE projects.

But at this scale, central actors and leaders of RE initiatives are in fact residents and especially farmers. These trends observed with a small number of case studies are confirmed on national scales. In Germany, 42% of RE production capacities are owned by citizens (Trend: research, 2017). The role of farmers in these local RE transitions, as initiators and actors, is absolutely central. That is the case in our field study, as well as at the national level. In Austria, for example, between 1983 and 1998, around 300 BDHs were implemented by farmer cooperatives (Rakos, 1998). In Germany, 10.5% of electric RE installed capacities are owned by farmers (small plants), rather than the "Big Four" energy companies, which only have 5.4% (Trend: research, 2017).

Different factors explain the central role of farmers in local RE transitions: there is in farming activity a specific relationship to nature, place and time, but also to risk and autonomy, in which more than elsewhere, RE makes sense. In addition, there is a technical culture that allows and conduces to innovation processes like the development of new types of RE plants. The organization of the agricultural sector also plays a role when it comes to developing a company or a cooperative to manage a collective RE plant. The importance of agricultural professional culture is, for example, raised by Wirth et al. (2013) to explain the unequal development of biogas plants in Austria despite identical incentives measures at national level (like feed-in tariffs). Agricultural guilds and professional culture constitute bridges and support the development of RE projects by farmers that then enroll inhabitants and municipalities. All these factors explain why more than other actors, farmers are implementing RE and play a central role in local RE transitions (Dobigny, 2015).

The field study thus reveals that local RE transition is conducted by unconventional players in the energy sector: individuals, farmers and local communities. Then how to understand this choice? What are their motives?

Motives of local RE transition

Different motives support the choice of local RE implementation. These collective RE projects have in common the support of social, political and ecological values and they are part of an empowerment process.

An empowerment process

First, the idea of collectively and locally doing-it-themselves, becoming an actor and deciding what and how energy could be locally developed, is clearly evident from interviews, as reflected by this interview extract with the Mayor of Freiamt:

> I still say that the inhabitants of Freiamt[1] are in fact free citizens, and wish to make decisions themselves, were highly important issues. The first aspect [of the local RE project] was therefore that we wanted to decide ourselves,

having a genuine identification with the projects, playing an active role within this sphere. The other reason was, of course, given where they live [the Black Forest], small farmers had low incomes and renting their land in this way, provided them with an additional income. This way, we were also ensuring that they could remain on their farms and that they did not have to leave. It is something that we have already seen in mountain regions, people leaving the area because it is not profitable, and this way we earn an income from doing nothing and that helps, of course. These were the two main aspects of the project.

Through local RE projects, there is an energy (re) appropriation, and hence an empowerment process, with both social and political dimensions. Thus, local energy autonomy is a gradual process. In most of the cases in the field studies, at first, one RE project was achieved and then it led to another RE project a few years later, and that step-by-step resulted in a 100% RE territory. Although this empowerment process is energy-based, it has an impact on many other matters at a local level (political, social, economic, etc.).

Supporting local stakeholders and the economy

Two aspects that the Mayor of Freiamt underlines in the previous interview extract are strong features across the field studies: on the one hand, the empowerment process and, on the other hand, the support to local stakeholders (especially farmers) and the backing of the local economy through RE projects. As mentioned before, the towns face numerous socioeconomic difficulties, and in that context, RE projects are seen as projects with positive impacts in terms of the local economy and employment. Supporting local stakeholders and especially farmers is a strong motive of RE projects in the four towns studied. Sometimes, like in Mureck, inhabitants provide support to local farmers and their RE project (a biomass district heating) despite the fact that the RE produced was more expensive than conventional energy at the beginning of the project:

> From 1995 to 2000, we were about a third more expensive compared to the oil or coal heating we used at the time. It was like that. Despite everything, Mureck, the inhabitants of Mureck agreed to support us. They said yes, we want that farmers, local people, do it, supply us with energy. Even though it costs us more, we want to help protect the environment and keep more money in the region. In addition, they obtained a secure supply, and it worked, half of the households were connected to the grid at the launch of the plant.
>
> (Farmer, leader and initiator of Mureck's RE projects)

Even though the financial advantage of the RE project facilitates the residents' support, it isn't always the case. Inhabitants could support a RE project that isn't economically advantageous, but matching their values, like in Mureck. Thus, the economic factor the first motive of local RE projects. That interview's extract underlines other reasons that are recurrent in the field study: supporting local stakeholders, responding to a local issue (socio-economic or environmental, like pollution) and the importance of oil shocks on representation, i.e., energy security issues.

Oil shocks indeed deeply marked an entire generation and widely contributed to the development of energy alternatives in the 1980s and 1990s to ensure a secure energy supply. As this interview's extract demonstrates, energy security is a concern for Mureck RE project initiators who belong to that generation. Adding to the apprehension of lacking in fossil energy supply, the growth of environmental concerns at the same time helped to develop more ecological alternatives, like RE. If these events had a direct influence in the 1980s, they left traces in the following decades, especially in the representation of this generation of actors that are engaged in later RE projects.

Ecological dimension

Then, the ecological dimension of the RE project is, of course, a motive. It is seldom the first, but it's a strong motive, as explained by this inhabitant of Jühnde, a member of the local energy cooperative:

> From the start, myself, I found the idea exciting because it was something innovative, it was from an ecological point of view, it really was something good, and then apparently it would financially not be a higher cost than the cost we had at the time while being heated with fuel … It had only positive aspects. So thinking about yourself, maybe thinking about your children and grandchildren, and thus to future generations, we said to ourselves, let's do something … good.

The ecological dimension of the RE project plays a role, sometimes with a moral dimension like this interview extract underlines (doing "*something good*"). Of course, the hierarchy of motives changes from one actor to another, and there is a wide range of motives in a community, like this interview's extract with the same inhabitant of Jühnde shows:

> The fact that 140 houses at the time no longer had a CO_2-producing chimney was still a very important aspect […] Yes, yes. Not everyone, for others it may have been the [positive] financial aspect in the future … For some, it must have been the little bit [of the] idealistic side of the project. And then, there is certainly a part of it that did it without great conviction at the beginning, simply because "Oh well, if the neighbors do it, I do it too". In fact, there is everything in a village like that! And then there will have been people who from the start said "no".

In a collective RE project, motives differ from one actor to another, and the support of the project is often a combination of different motives, with changing priorities. It could be, for example, the easier usage that a collective energy system allows, like in the case in Jühnde, where there is no more need for an individual boiler with the biomass district heating (i.e., no more individual maintenance and relative costs). Cheaper energy and/or a stable energy price are also arguments in favor of RE projects. In most cases, RE produced is cheaper, but sometimes at the beginning of the project, the energy costs would be the same (Jühnde) or more expensive (Mureck). But with the increasing costs of fossil energy, RE locally produced became cheaper in studied towns, when it wasn't the case at the beginning (Table 10.1).

Table 10.1 Key financial data: the case of Jühnde

Costs (biogas power plant implementation)
5.4 million euros
Financing
Own resources: around 0.5 million euros
Subventions (EU funds): around 1.5 million euros
Bank loan: around 3.5 million euros
Governance
Cooperative (195 members: heat consumers, farmers, municipality, the church and some external members)
Estimated saving of CO_2 equivalent
Around 3,500 t/year
Cost savings
An average household in Jühnde saves around 750 euros/year through the use of bioenergy (heating and domestic hot water. Data from 2014 in comparison with the costs of fossil fuels)

Sources: Weglarz, Winkowska and Wojciech (2015); Ruwisch (2005).

Social, fair and democratic values

Another important motive in favor of the local RE project is its social and democratic dimensions. For the Mayor of Jühnde (during the project), it was a strong motive:

> I think what is very important is that this plant belongs to the citizens of here, to the members of the cooperative. So not to the town … to the Mayor or someone else, but to the people who are members, we have 194 members in the cooperative and they are owners. And everyone has a voice. Whether he is rich or poor, he has a voice. A voice at the table here. And I think it's a social reflection. So not: the one who has money has more say, but everyone here has a voice and can take the floor and that's socially just. I think that's it … that was what was close to my heart […] This is also a social reflection, the fact that even people who have little money can participate […] It is also an extremely important thing for peace in the village […] And I believe that it is a necessary condition, to pay attention to that, to make

sure that the balance is respected. Let it not always be those who earn the most who have an advantage, but also every normal citizen. For me, it's always a bit in the foreground. Of course, profitability must also be ensured, it is very clear.

With an initial deposit made accessible to the greatest number of people and an organization in which each member has one voice, regardless of the number of shares he or she owns, the fairness of the cooperative's model could really be a strong motive in favor of local energy production. There are indeed social, democratic and fair dimensions in local RE projects through its governance (mostly citizens based), legal form (cooperative, citizen company, public-private company, public plants, association, etc.), energy price (cheaper), local profitability (local economy), management (empowerment) and so on. According to the Mayor of Güssing (that initiated and led the RE transition of the town), reaching cheaper energy was a fair and social aim in order to avoid poverty. Güssing has developed a model of mixed RE companies (public-private) with a municipal majority of shares (the remaining part of it is owned by citizens or local investors) to avoid potential speculation on the energy price and guarantee low energy prices.

Local and citizen-based RE projects then support ideals ("socially just", "something good", "the little bit idealistic side of the project", "social reflection") with social and democratic values that promote a fairer social organization and citizen empowerment.

Thus, local and collective RE implementation is based on social, political and ecological values. This energy (re)appropriation inscribes itself in an empowerment process, with social and political implications. Supporting local stakeholders is therefore a strong motive, as well as the ecological, social, fair and democratic dimensions of these RE projects.

Social and economic implications

Economic benefits and energy tourism

Local RE projects have indeed social and economic spillover effects for the territory. First, developing RE at a local stage has an impact on employment that could be direct (e.g., RE plant management) or indirect (e.g., conservation of local business).

Thus, one unexpected result of local RE implementation is the development of energy tourism. While the implementation of large RE plants, like large wind turbines farms, may have a negative effect on local tourism activity according to some authors (Lilley, Firestone and Kempton 2010; Riddington, et al., 2010; Broekel and Alfken, 2015), others reported no significant negative impacts from wind turbines on local tourism and demonstrated that new RE energy facilities could represent attractions for a certain kind of tourist (Aitchison, 2004; Frantál and Kunc, 2011; Frantál and Urbánková, 2014, Lilley, Firestone and Kempton 2010). In the case of local RE transition, we observed a positive effect on tourism demand. As these towns were among the first to reach a local RE autonomy, they attracted people all around the world (Europe, America, Asia, etc.) interested in visiting RE plants and to find out more about implementation processes, positives and difficulties, governance and so on. Visitors were community leaders, members of municipal councils, energy administrations or institutions, associations, NGOs and so on. As a result, there were between 600 and 1,000 visitors a week in Güssing and between one and three groups a day in Jühnde in 2008, while there was no tourism activity before the RE transition. In Jühnde, a small town with fewer than 800 residents, inhabitants organized themselves to respond to the demand and volunteers were trained to conduct visits. That created a complementary activity for these guides.

In Güssing, due to the number of visitors, tourism activity, also named "eco-energy tourism" (David et al., 2019; Jiricka et al., 2010) has had a huge impact on the local economy and employment (restaurant, hotel, transport, etc.) and is well organized and developed. A RE research center (the European Centre for Renewable Energy) was implemented in 1996 and more than 30 demonstration RE plants using different technologies (biomass, wood, biogas, solar, photovoltaic) can be visited. Visitors are then "expert-oriented energy tourists" (community leaders, enterprises, universities, technical schools, etc.) defined by Jiricka et al. (2010, p.58) as "professional visitors, who visit to gain new knowledge for their communities, their enterprises, or their business". They are interested in technical innovation, management and implementation processes, transfer of technical knowledge and the possible adaptability for their own region (Jiricka et al., 2010). Moreover, the municipality has developed a tourism offer to encourage people interested in energy issues to stay longer in the area, for example, with the construction of new hotels and infrastructure (Zotz, 2008).

In Freiamt and Mureck, energy tourism is adding to existing tourism activity. Indeed, the small size of RE plants at a local stage does not have a negative impact on traditional tourism (leisure and family). Even in the case of Freiamt, located in the Black Forest with established hiking tourism, the six wind turbines disseminated on hills (i.e., visible) have not had a negative impact on the number of hikers. Rather, wind turbines appear on the tourism and hiking map of Freiamt. A municipal staff manages energy tourism (RE plants visit offer). In 2008, 20,000 tourists visited Freiamt for whatever reason and 6,000 energy tourists per year were estimated in Mureck (SEEG, 2008). In both towns, energy tourism has reinforced existing activities and infrastructure (like hotels and restaurants) and developed new ones (especially for farmers or citizen groups that show their RE plants).

Therefore, has this tourism activity been limited to a couple of years, or has it lasted throughout the pioneer period and diminished when local RE transition became more common? We can suppose that tourism attractiveness decreases with the generalization of RE transition. Nevertheless, three factors contribute to the durability of this tourism activity: first, the fame gained by these towns related to their RE autonomy and was disseminated by numerous press articles and TV reports; second, the development of tourism infrastructure, interests and leisure activities. In fact, there were 1,700 overnight stays in Güssing in 1990, 25,000 in 2007 (the boom period of energy tourism) and today there are 12,890 (2019). Although overnight stays have diminished in the town after a boom period, it is not the case for the Güssing's region, where overnight stays continue to grow – from 27,000 in 1990, to 250,000 in 2007 and 310,258 in 2019 (Zotz, 2008; Wurglits, 2020; Statistik Austria). Thus, fame and tourism continue to benefit the territories, even if energy tourism is less important. Moreover, if the number of overnight stays could indicate a trend, it doesn't exactly reflect energy tourism, which includes a large share of one-day visitors. Third, apart from these "expert-oriented tourism" (experts, enthusiast people or businesses), energy companies, RE developers or municipalities are increasingly developing an "experience-oriented tourism" (leisure-oriented) to attract additional segments of tourists, such as young people, families, or seniors (Jiricka et al. 2010; Frantál and Urbánková, 2014). This objective is, for example, reached by the town of Güssing, which has developed, alongside tourism infrastructure, tourism attractions like sports events around RE plants (e.g., the Eco Energy Marathon Güssing), a 125-km bicycle trail in the region or a holiday camp around an RE theme. Examples of sport or cultural events around RE plants are numerous (e.g., the Dragon Kite Festivals under wind turbines in the Czech Republic, Frantál and Urbánková, 2014) and we can suppose that even if local RE transition becomes more common, the continuity of such events will ensure durable tourism activity. RE could also generate tourism interest in the long term, as supported by the publication in Germany of the first renewable energy tourism travel guidebooks (Frey, 2010; Frey, 2014). These different tourism developments around energy in numerous countries worldwide make energy tourism an emerging field of study (Frantál and Urbánková, 2014; Jiricka et al. 2010; David et al. 2019, Vourdoubas 2020).

Energy tourism could have a huge economic impact in these towns, both direct and indirect. The development of new businesses in the energy sector, like RE and eco-material businesses or RE research centers, also creates new employment. In addition, due to the fame of the town, the positive and ecological image associated with the town or a cheaper energy price (like in Güssing) with a local RE transition could lead to the establishment of new companies. Thus, more than 50 new enterprises have been established in Güssing since its RE transition. As a result, from 1998 to 2008, 1,100 new jobs were created in Güssing, a town of 4,000 inhabitants. In the other towns studied, the number of new jobs created was less spectacular, but in all towns, local RE transition has had a positive economic impact and new jobs were created.

New solidarities

Apart from these positive economic effects, local RE transition also has social implications such as new solidarities both within and outside the community. Due to these local projects and the involvement of local residents, they have had occasions to meet up, work together and get to know each other. Although they are little towns, with a low number of inhabitants, people did not necessarily know their neighbors. RE projects, therefore, created opportunities to meet and contribute to new relationships within the community, as explained by this inhabitant of Jühnde, living in the village of 800 inhabitants for more than 20 years who only knew three or four people beforehand:

> I would say people are closer to each other. We ourselves have many more contacts [...] with a lot of people from the village, with whom we worked, with whom we were often, with whom we often spoke ... it brought people closer to each other. Enormously. Myself, I would now speak to almost everyone in the village knowing who he is, the person knowing who I am, and without any embarrassment, without any distance [...] We're a bit like a big family now, we almost all address each other using the "tu" form [laughs]! Oh yes it's true! We are ... this is our project. We are a member, we are a member of the cooperative, we are colleagues, we are ... We are more than neighbors, we have common interests. And common ideals.

Due to local RE projects, new solidarities were thus created within communities, between inhabitants, but also outside. These solidarities could take the form of interpersonal relationships, as well as cooperation and exchange around energy. Inhabitants could make (private) wood donations to the municipal biomass heating district, for example. Therefore, energy autonomy does not make an autarky. Many relationships are created between inhabitants and people outside the towns due to energy tourism – exchanges of knowledge and skills, international twin town programs around RE transition, local cross-border RE projects and so on. The inhabitants, and especially farmers that usually can't easily leave their farm to travel, appreciate this openness to the world.

Recognition, identification and collective identity

Beyond economic benefits and new solidarities, local RE transition also has symbolic and identity implications. Energy tourism, town mediatization and the image of a "model" town of RE transition contribute to a sense of pride and recognition of inhabitants. In addition to tourism and media coverage, recognition is also expressed by the prizes and awards won by towns. Mureck, for example, was awarded the World Energy Globe in 2001, a first for Austria, the European Solar Prize in 2006 (*Europäischen Solarpreis*), the title of the most innovative village of Austria in 2007 (*Innovativste Gemeinde Österreichs*) – leading to the Federal President of Austria visit in July 2008 – and the Austria Solar Prize in 2010 (*Österreich Solarpreis*). Güssing was also awarded, in 2004, the distinction of the most innovative village of Austria, the Austria Solar Prize and the European Solar Prize, and in 2005, the Energy Globe Austria. These prizes are proudly displayed on plaques located at the city entry. All of this – tourism, mediatization, awards, visits of important politicians – contribute to the recognition of the inhabitants.

Moreover, RE plant appropriation and identification are strong in all towns studied. Indeed, inhabitants led and invested themselves and their time in these projects, sometimes for several years, as explained by an inhabitant of Jühnde:

> Ah, we are proud [of the RE plant]. I even put "bioenergy village" on our private cards, our business cards! We're proud, we've spent a lot of time there too, we've been working on it for five years, eh. We spent hours there. It's normal that we identify a little bit with it.

Inhabitants and stakeholders are proud of having successfully carried out these projects, collectively, that contribute to RE plant identification and then to the town. She carries on:

> We feel strong. One feels Jühnde. We feel like a resident of Jühnde and not just individuals […] Now we are "Bioenergiedorf" [the bioenergy village], we identify to that.

Around RE projects, an individual and collective identity has formed, while before RE transition, identification with the town was low, as underlined by the Mayor of Güssing:

> The identification [of the inhabitants] with the town is great. It hasn't always been that way. Pride is adding to that: we live in a model city for renewable energies. It's a source of pride.

Thus, local RE transition has positive implications in terms of solidarities and collective identity. These social benefits generated by community-based renewables are also observed by Süsser and Kannen (2017) in pioneer communities of North Frisia (Germany) where RE development has increased community spirit and cohesiveness between community members, as well as social stability and social diversity.

But one interesting fact is that identification does not only concern people engaged in RE projects. In most towns studied, people not engaged in local RE projects also identify with it. While the local ownership of the RE plant is often presented as a necessary condition of local acceptability and identification (among others, Süsser and Kannen, 2017), it isn't always the case. Our field study raises a lot of examples in which citizens identify themselves with RE projects in which they are not engaged or there are no community-based RE plants (large wind turbine farms of investors, for example). How is this understood? In fact, what changes in these rural communities with RE transition is the town's image and then the inhabitants' image, as described by the Mayor of Freaimt:

– What does this energy autonomy represent for you?

– Well, I think it has played a very big role in the self-confidence of our population, because with this energy autonomy, a … yes, respect for the population rose. We are a little bit off the line, it is always "ah those, in the depths of their countryside", so rather negative, viewed by the townspeople and today the townspeople come and look at what has been done in the countryside. So it played a huge role in the self-confidence of our population. Pecuniary, it doesn't matter at all because the municipality no longer has taxes [with this RE project]. But indirectly, what I said earlier, is that people just stay here. So the significance is actually more … what do you call that? … in soft factors rather than hard factors. So the significance is not that municipality has a higher income, the significance is that our citizens have more income, that they stay here - this soft factor of location - and I would like to say, psychologically, this self-esteem. This has risen sharply in recent years. So the people here were a bit … shy or, not shy, they were directly seen as "country people". And that has changed. People are now saying "we are proud" and "we are happy" and "we have set an example for you". And that's just the way it is, it's very important for all other political projects that you naturally have!

With local RE transition, towns demonstrate that innovation does not only happen in urban areas. It links to a change in the town's image, and thus, the image that inhabitants have of themselves. Thus, local acceptability is less linked to ownership, and identification is less related to the personal participation *than* to the modification of the town's image (e.g., "most innovative village", "model town", "pioneer village", etc.), and thus the inhabitants' image, which constitutes a new collective identity (for all citizens, engaged or not in RE projects).

Conclusion

Local RE transition generates economic and social benefits, such as positive economic impacts, employment and energy tourism, new solidarities within and outside communities, identification and collective identity. Unconventional players in the energy sector lead this RE transition: mostly citizens, farmers and municipalities, with social, political and ecological motives. Indeed, local energy transition is part of an empowerment process, with social, economic and symbolic implications. It also implies that technology isn't neutral. Energy and technical choices (small RE plants, local technical systems and the energy grid) have social, economic, political and symbolic implications.

Due to the time gap between RE projects studied and the United Nations Sustainable Development Goal (SDG) set in 2015, communities didn't explicitly target them. Nevertheless, SDG 7 (i.e., ensure access to affordable, reliable, sustainable and modern energy for all) and SDG 11 (i.e., make cities and human settlements inclusive, safe, resilient and sustainable) are actually reached by the 100% RE territories studied.

Challenges to local RE transition could be the financial viability of community-based RE projects or local opposition. They are not discussed in this chapter, as they were not prominent in this field study. The absence of actual local opposition could be explained by economic, social and symbolic benefits in these towns. Long-term financial viability is ensured by the local consumption of energy production. If a broad range of consumers is targeted, the financial balance could be weakened by national or international energy markets (like the oil mill of Mureck producing bio-fuels). Only targeting local energy consumption is one of the success factors and the long-term financial viability of locally based RE projects. Building local RE projects together is another. Multi-actor cooperation, that is, inhabitants, farmers, local stakeholders and municipalities in the same RE project, is a key factor to the success of local RE transitions in Austria and Germany. This cooperation contributes to a shared vision of the local energy transition that is central to being locally supported. Local opposition results in a lack of a shared vision of local energy transition.

Note

1 "Frei" means "free" in German.

References

Aitchison, C. (2004) *Fullabrook wind farm proposal, North Devon: Evidence gathering of the impact of wind farms on visitor numbers and tourist experience*. Bristol: University of the West of England/Devon Wind Power.

Broekel, T. and Alfken, C. (2015) 'Gone with the wind? The impact of wind turbines on tourism demand', *Energy Policy*, 86, pp. 506–519.

David, L.D., Molnar, C., Kosmaczewska, J., Fodor, G., Zsarnoczky, M., Varga, I. and Palencikova, Z. (2019) 'Ecoenergy tourism, study into some aspects of relationship between use of renewable energy resources and sustainable regional and rural development', in *Proceedings of the 18th International Scientific Conference 'Engineering for Rural Development'*, Jelgava, Latvia, 22–24 May 2019. Latvia University of Life Sciences and Technologies, pp. 1478–1483.

Dobigny, L. (2019) 'Sociotechnical morphologies of rural energy autonomy in Germany, Austria and France', in Lopez, F., Pellegrino, M. and Coutard, O. (eds) *Local energy autonomy. Spaces, scales, politics*. London: ISTE and WILEY, pp. 185–212.

Dobigny, L. (2015) 'Le rôle central des agriculteurs dans les projets d'EnR. Apports pour une socio-anthropologie des énergies renouvelables', in Zélem, M.-C. and Beslay, C. (eds) *Sociologie de l'énergie*. Paris: CNRS Éditions, pp. 349–356.

Evrard, A. (2013) *Contre vents et marées, Politiques des énergies renouvelables en Europe*. Paris: Presses de Sciences Po.

Frantal, B. and Kunc, J. (2011) 'Wind turbines in tourism landscapes: Czech Experience', *Annals of Tourism Research*, 38, pp. 499–519.

Frantál, B. and Urbánková, R. (2014) 'Energy tourism: An emerging field of study', *Current Issues in Tourism*, 20 (13), pp. 1395–1412.

Frey, M. (2010) *Deutschland – Erneuerbare Energien entdecken*. Baedeker Reiseführer. Berlin: Mairdumont Verlag.

Frey, M. (2014) *Deutschland – Erneuerbare Energien erleben*. Baedeker Reiseführer. Berlin: Mairdumont Verlag.

Jiricka, A., Salak, B., Eder, R., Arnberger, A. and Pröbstl, U. (2010) 'Energetic tourism: Exploring the experience quality of renewable energies as a new sustainable tourism market', *WIT Transactions on Ecology and the Environment*, 139, pp. 55–68.

Lilley, M.B., Firestone, J. and Kempton, W. (2010) 'The effect of wind power installations on coastal tourism', *Energies*, 3(1), pp. 1–22.

Rakos, C. (1998) *Lessons learned from the introduction of biomass district heating in Austria, E.V.A.* Vienna: Austrian Energy Agency.

Riddington, G., McArthur, D., Harrison, T. and Gibson, H. (2010) 'Assessing the economic impact of wind farms on tourism in Scotland: GIS, surveys and policy outcomes', *International Journal of Tourism Research*, 12(3), pp. 237–252.

Ruwisch, V. (2005) 'Wärme- und Stromversorgung durch heimische Biomasse. Bioenergiedörfer - Bausteine einer nachhaltigen Energieversorgung', in *Conference presentation*, IZNE, Offene Universität, October.

SEEG (2008) *Bioenergie-kreislauf Mureck*. (SEEG documentation).

Statistik Austria (2020) http://www.statistik.at

Süsser, D. and Kannen, A. (2017) ''Renewables? Yes, please!': Perceptions and assessment of community transition induced by renewable-energy projects in North Frisia', *Sustainability Science*, 12, pp. 563–578.

Trend: Research (2017) *Eigentümerstruktur: Erneuerbare Energien. Entwicklung der Akteursvielfalt, Rolle der Energieversorger*, Ausblick bis 2020. (Trend: Research Report).

Vourdoubas, J. (2020) 'The Nexus between tourism and renewable energy resources in the Island of Crete, Greece', *American Scientific Research Journal for Engineering, Technology, and Sciences*, 63, pp. 28–40.

Węglarz, A., Winkowska, E. and Wojciech, W. (2015) *A low-emission economy starts with municipalities*. Berlin: Adelphi Research.

Wirth, S., Markard, J., Truffer, B. and Rohracher, H. (2013) 'Informal institutions matter: Professional culture and the development of biogas technology', *Environmental Innovation and Societal Transitions*, 8, pp. 20–41.

Wurglits, M. (2020) 'Tourismusbilanz 2019: Viel Licht, viel Schatten', *Bezirksblätter*, Güssing/Jennersdorf, 5, February.

Zotz, A. (2008) 'Ökoenergie-Tourismus in Güssing', *Integra*, 1, pp. 18–19.

Part V

Technological issues in energy transition: market, grids and smart cities

11 Digital and energy transition in French cities: limits and asymptote effects

Raphaël Languillon-Aussel

Introduction: energy, the neglected *alpha* and *omega* of the French smart city?

The smart city is not truly a French urban model. The idea first emerged in the United States in the mid-2000s, with no direct connection to the issue of energy, or even sustainability. The initial idea of the smart city is rather based on the observation of two ongoing revolutions: that of the rapid urbanization of the world, on the one hand, and that of the digital industry, on the other (sometimes called the third or even the fourth industrial revolution). In 2005, President Bill Clinton asked one of his national digital champions, Cisco, to establish an industrial bridge between the two revolutions by formalizing the smart city model. In 2008, IBM formalized its "smarter city" proposal on the principle of an industrial project that linked the growing urban market to the digital one.

The year 2008 is a key moment in the emergence of the smart city model for several reasons (Courmont and Le Galès, 2019). According to UN Habitat, the world's urban population exceeded the world's rural population that year, reflecting the worldwide urban revolution (UN Habitat, 2009). Also in 2008, the first smartphones appeared, with the Apple iPhone, which turned people into mobile sensors – used in the smart city. The subprime crisis, which also struck in 2008, led to deep political and economic transformations in the urban fabric and its governance, opening a window of opportunity to the digital giants who rushed into it with their various proposals for smart cities.

Despite the proliferation of smart city initiatives and projects in France, Europe and the world, and despite the age of the model (around 15 years in 2020), its definition is still unclear. As the question of the model is still being debated, the terminology has also not stabilized. Japan uses the term "smart community" as such (Languillon-Aussel, Leprêtre and Garnier, 2016). In France, we find both terms, although that of "smart city" is more widespread. Michael Batty defines the "smart city" as being a city structured by the genesis, collection, management, processing and almost instantaneous analysis of big data, corresponding to digital data generated by the dissemination of sensors connected in urban spaces and, in particular, in networks (Batty, 2013).

DOI: 10.4324/9781003025962-17

Energy considerations are therefore of secondary importance, compared with digital considerations, in the definition of the smart city in France as elsewhere in the world. As such, that the relationship between the smart city and the sustainable city – the latter being much more focused on energy – cannot go without saying. What can be said about the smart city regarding energy issues? How are the French energy players positioned in the formalization of the smart city? Is the smart paradigm compatible with the energy transition of urban territories in France and Japan, and the achievement of objectives 7 and 11 of the Sustainable Development Goals (SDGs)?

With the appearance in the United States of the smart city model in the mid-2000s and its dissemination in France at the turn of the 2010s, the energy issue has become the *alpha* and *omega* of cities and urban territories. On the one hand, the smart city aims to rationalize the consumption of resources, of which energy is a part – in particular, electricity. On the other hand, the myriad sensors and digital terminals on which the operation and optimization of the smart city depend requires large amounts of energy, particularly electricity, to function. Electricity is thus both the stake and the condition for the functioning of smart cities. In this, the smart city meets the energy challenge both with regard to objective 11 of the SDGs ("make cities and human settlements inclusive, safe, resilient and sustainable"), which is its omega (the development issue), and objective 7 ("ensure access to affordable, reliable, sustainable and modern energy for all"), which is its alpha (the operating condition).

The actors of energy and digital tech, and their urban strategies in France

The energy issue was central in the European model of the sustainable city, as defined after the Rio Earth Summit in the early 1990s. The urban sustainability model was then generalized by Agenda 21 of the European Union, which has been translated locally by countless local Agenda 21s in France, as in other European member states. In the mid-2000s, the sustainable city was a model that had, however, already lived. When the smart city emerged in 2005, 15 years had passed since the implementation of local Agenda 21s, corresponding to a generation of political decision-makers. The smart city is positioned on a different level than the sustainable city, partly because of the new players who carry it.

The newcomers of city making and governance

Most of the time, researchers working on urban models underestimate the role of the succession of generations of decision-makers in the course of the models' life cycle (formalization, test application, dissemination of the model, then decline and disappearance). However, urban models are effectively formalized and undertaken by institutional actors who are themselves embodied by individuals, decision-makers and investors, who do not stay more than 15 years in positions of responsibility that allow them to make decisions affecting the urban fabric – positions occupied, most of the time, at the end of one's career. This generational effect is verified by the sustainable city that appeared in the early 1990s, the smart city, which appeared in the mid-2000s, and even the safe city that emerged at the turn of the 2020s, in each case defining a 15-year cycle.

In addition, these cycle effects affect not only decision-making individuals but also the institutional actors themselves. Thus, the proliferation of smart city development projects can be explained by the arrival of new institutional players in urban construction and governance, who differ from those who have tradition-ally approached the city as their core business, such as real estate developers (Baraud-Serfaty, 2013). Since the 1980s, we have thus observed three waves of new entrants into the making of urban spaces:

1. Since the 1980s, the financialization of real estate has given rise to new play-ers in the urban fabric from the finance sector (such as listed real estate com-panies). Their positioning vis-à-vis the urban area is characterized by new, sometimes speculative practices, such as real estate securitization. The sub-prime crisis of 2008 was essentially the result of this financialization.
2. In the 1990s and early 2000s, the actors of sustainability arrived, often from the energy sectors (such as Siemens or General Electric), following the pro-motion, by the public authorities, of the sustainable city model. In France, these are players such as Engie, Suez and Veolia, which were taking on an unprecedented leadership role in so-called sustainable urban development projects. This phenomenon was concomitant with reflections on the deregu-lation of the energy market in Europe.
3. Finally, the years 2000 and 2010 saw the rise in importance of what Isabelle Baraud-Serfaty calls the "urban players of tomorrow" (Baraud-Serfaty, 2013). These actors, resulting from new information and communication technolo-gies, and then from digital technologies, disrupted the political balances and business models of the making and governance of cities: for example, com-panies like IBM, Cisco, Toshiba, Samsung, Uber, Amazon or Google. In France, we can identify three sub-categories of digital players: the internet giants (internet service providers, such as Orange or SFR, or browsers, such as Google or Ecosia), digital equipment and software manufacturers (such as IBM, Toshiba, Siemens) and digital platforms (such as Uber, Deliveroo, Air BnB, etc.). All these actors have both a vision and political objectives and even, in the case of some platforms, hegemonic and oligopolistic claims.[1]

Each of these waves added to the players in the previous ones without eliminating them but leading to changes in partnerships and business models. Most of the cur-rent smart city projects involve a large number of players from each of these three waves. In Lyon Confluence, for example, Bouygues Immobilier ("pure urban"), EDF (energy) and Toshiba (digital) are in charge of the development of an experi-mental smart community site called Hikari.

In France, however, the absence of national champions in the digital sector has pushed large French groups that already exist to leave their core business (mainly urban planning) to experiment with smart city projects themselves, in partnership with French telecommunications groups, like Orange, or with European, American, Japanese or Chinese digital companies. In this context, groups from the real estate sector, such as VINCI (which is developing a major start-up incubator specializing in artificial intelligence, named Léonard) or groups from the energy sector, are investing in smart technologies. In the smart OnDijon initiative in Dijon, for example, the leading company is Bouygues Énergies & Services (a subsidiary of Bouygues Construction), in conjunction with Citelum (a subsidiary of the EDF group), SUEZ and Capgemini. As for the smart city in Nice, energy players are at the heart of a consortium led by Enedis, with GRDF, EDF, Engie, GE, Socomec and the Nice Côte d'Azur Metropolis.

The new governance of urban planning in French smart cities

With the smart turn of the city, new companies from the information and communication technology sector are breaking into the making and governance of French urban environments. This is particularly the case for network planning and governance, including those for energy production and distribution. These new actors then modify the power relations, not only within urban governance, but also in the way of producing French cities. This power shift is also observable in public administrations, with increasing importance assigned to digital and IT sections in power relations within organizational charts and municipal, departmental or regional teams (Eveno, 1997). In terms of professional skills, the emergence of the "smart" has shifted the field of urban production from town planners to computer scientists (Picon, 2015).

The "smart" players thus penetrate the development and management of networks through the optimization of the management of various urban services: electricity, gas, water and sanitation, waste and urban heating networks, as well as transportation (Dupuy, 2014). Thus, for example, the production, distribution and consumption of energy are then optimized by the use of smart grids which, thanks to computer technologies, adjust the flow of electricity according to the information provided through feedback loops: digital players are then positioning themselves in the energy sector via the sector of big data and information systems.

The urban smart turn is thus accompanying transformations in stakeholder relations and modes of governance. In Dijon, the main objective of the smart city project is thus to control urban information systems and combine them into a single centralized system, represented by a control center – a new space that reactivates the imagination of the control room of cybernetics inherited from the 1960s. In doing so, the centralization of urban information systems aims, among other things, to circumvent the gatekeeping behavior of certain department heads who sometimes tend to jealously guard control of the information managed by their departments. This desire to put an end to the silo organization of the administration via the centralization of urban information systems aims to rupture or disrupt the excessive power of certain services within the municipal and inter-municipal administration of cities.

Urban governance and administration are not the only levels of power in France that have been transformed. The city's production methods have also evolved a lot, particularly through the generalization of in vivo experimentation. The use of experimentation is not new in the conduct of innovation policies. This practice most often takes place in protected and time-limited "niches" to ensure the technological functioning and viability of an economic model in the face of heavy investment (Erlinghagen and Markard, 2012). However, we observe an increasingly recurrent use of this instrument of public action in the conduct of environmental and energy policies since the 2000s in France, as well as in Europe.

This "government through experimentation" accords an important place to the interaction of socio-technical systems with the population to point out any flaws or resistance, while identifying "good practices" to be replicated. The proliferation of smart city initiatives raises questions about the generalization of systems, in terms of the scale of the city or of a larger territory, or by contrast, of a fragmentation that would tend toward a hybridization between technical macro-system and micro-grids. Such fragmentation leads to a double problem.

In the energy sector, on the one hand, this fragmentation is a technological and political challenge at different levels – national, regional and urban – due to the importance of the centralized French macro-system. This macro-system is difficult to circumvent because the weight of nuclear production requires a highly centralized governance. Bypassing this centralized technical macro-system by multiplying nationally decentralized, but locally re-centralized, urban micro-grids is a political and technological challenge. From the point of view of urban functioning, on the other hand, the proliferation of experiments, usually private, leads to fragmenting the city and bypassing the public governance of urban development by countless legal exception regimes.

Limits and asymptote effects of the digital turn of "urban" energy in France

French smart cities are often presented as sustainable cities, enhanced by digital technologies, as is the case, for example, in Nice smart city. In this sense, there seems to be a digital necessity in the energy transition of the sustainable city. As such, the multiplication of sources of so-called sustainable electricity production – solar panels and urban wind turbines, for example – leads to an increase in the sources of tension in technical systems; the production of renewable electricity, by nature, fluctuates according to climatic conditions (sunshine and wind, in the first place), which generate fluctuations in networks that can lead to a blackout if digital technologies fail to offer support by modulating and streamlining the systems. The energy sustainability of French cities, encouraged by local Agenda 21s in the 1990s and 2000s, made the deployment of urban smart grids compulsory at the turn of the 2010s.

Many studies that have worked to synthesize the definitions of smart cities agree that the achievement of urban sustainability is not limited to improving infrastructure, as it is also about encouraging residents to adopt more sustainable lifestyles by modifying their behavior relating to mobility, energy use and waste recycling. A smart city then has the potential to be a sustainable city, integrating the objectives of sustainable development by optimizing its operation (Eveno, 2014), which can then lead to the accomplishment of objectives 7 and 11 of the SDGs. Nevertheless, many studies based on social practice theories have recently highlighted the difficulty of predicting the impact of introducing new technologies into the domestic space (Gram-Hanssen, 2008); new technologies that run the risk of increasing energy-intensive standards of comfort (Shove, 2010).

As such, the energy sustainability of French and European smart cities may encounter four "progress asymptotes". First, the inhabitant asymptote, particularly acute in France, is twofold. On the one hand, individuals' resistance to the deployment of digital technologies in urban or domestic infrastructure is very important – as was the case with resistance to the introduction by Enedis of Linky, a smart electricity meter, in private homes. On the other hand, inhabitants and users' resistance to cycles of obsolescence and replacement of technologies that are too rapid, relative to their cognitive capacities to adapt to change, is also a factor of high importance. In addition to the increasingly shorter life cycles of digital technologies used in smart cities, the French population is aging rapidly, leading to a scissor effect between the speed of technological change and the cognitive capacity of the aging population to adapt to digital innovations.

Another asymptote of progress is that of open access and common use of private interconnected energy facilities. The principle is that, with the interconnection of renewable energy production units, the use, by all, of private equipment makes necessary not only the tacit acceptance of common use on the part of the owners but also a complex compensation system to which the French systems have not yet found a satisfactory answer – unlike Japanese players like Toshiba, which integrate point systems. Finally, the issue of energy saving quickly encounters counterproductive effects, particularly with the rebound effects of energy consumption permitted by the proliferation of electricity-hungry digital terminals, which tend to cancel out the gains in consumption enabled by the smart city.

Japanese smart communities and French smart cities: cross-cultural experiments through urban energy innovations

In France, the proliferation of smart city projects is, above all, the result of collaboration between local authorities – often inter-municipal or metropolitan bodies – and consortia of private companies, in which the leading players come from the sectors of construction, telecommunications or energy. In Japan, the impetus rather comes from the state, which plays a major role in the emergence of the Japanese "smart" turn, due to its specific "developmentalist" approach.[2] Government agencies also help Japanese conglomerates (the so-called *keiretsu*) to export their models abroad. France is, in this context, a target as much as a privileged partner. Can Japanese smart communities serve as a model for French smart cities?

What is a Japanese smart community?

Closely following American initiatives, Japan formalized an industrial proposal for smart cities – calling them smart communities – at the end of the 2000s (Languillon-Aussel, Leprêtre and Granier, 2016). From 2009, the Japanese government established research groups made up of academics and major Japanese firms to determine which smart technologies should be developed as a priority. It retained 26 technologies in January 2010, which involve a massive deployment of renewable energies, the establishment of energy management systems (Energy Management System or EMS) and the introduction of new generation transport systems (Leprêtre, 2018).

From 2010, the Ministry of Economy, Trade and Industry (METI) began implementing several programs aimed at developing smart urban industrial products at local levels. The most important is called the "Next Generation Energy and Social System".[3] Launched in 2010 for five years, it had a total budget of 40 billion yen – over 320 million euros. Following a call for projects, to which more than 20 territories and consortia responded, four "smart communities" were selected: the city of Yokohama for large urban smart grids, the city of Toyota for transportation, the Kyoto prefecture for its project within the scientific city of Keihanna and an industrial reconversion experiment in Kitakyushu focusing on hydrogen.

The disparity of the four cases makes it possible to test various elements and innovations, without considering all the complexity of the "city" objects. The experiments thus concerned small spaces: 225 households in Toyota, 230 in Kitakyushu, 700 in Kyoto Keihanna and 4,000 in Yokohama, which is by far the largest initiative in the country. Each initiative was led by a local community (the municipality, or the Kyoto prefecture, in the case of Keihanna) in association with a consortium of private companies.

The national labeling program does not exhaust the diversity of smart community experiments in Japan. Completely private projects, such as Fujisawa Sustainable Smart Town (undertaken by Panasonic on one of its brownfield sites, west of Yokohama), or mixed projects, such as Kashiwa no Ha (led by a private consortium, the leading company of which is a real estate firm, Mitsui Real Estate, in partnership with public universities and the prefecture of Chiba, northeast of Tokyo), allow testing of a wide range of technologies in different kinds of territories, from core urban to suburban and rural ones (Languillon-Aussel, 2018).

Initially centered on the development of competitive technologies, with a view to international deployment and the reduction of greenhouse gas emissions, the "smart communities" program underwent a significant reorientation of its objectives after the accident at the Fukushima nuclear power plant in March 2011, which led to the shutdown of nuclear production within the Japanese archipelago. The fragility of the Japanese electricity grid that revealed this traumatic event prompted public and private actors to prioritize the development of energy management systems and incentives to reduce consumption during peak periods. The Fukushima accident thus put the issue of energy policies at the very top of the agenda of smart communities (Languillon-Aussel, 2018).

For example, in Kashiwa no Ha, the centralized big data management system, known as AEMS (Area Energy Management System), has assumed new socio-economic meaning in the post-Fukushima context, promoting the decentralized management and distribution of energy. The Fukushima nuclear accident, in fact, revealed two major weaknesses in civilian nuclear power: the permanent risk inherent in controlling fission and the risk of an energy blackout inherent in centralized national or regional distribution systems. The AEMS system was developed in response to these two risks, and it acquired a particular social and economic impact after the disaster with the population, investors and economic actors. In this sense, the empowerment of energy production and distribution in small, decentralized units appears to be a response to the challenges of a concentrated national energy park, vulnerable to the vagaries of the nuclear sector. The AEMS-based smart city has become a post-Fukushima urban model of energy resilience.[4]

The Japanese partnership in French smart community projects: the example of the Lyon smart community

Japan has not restricted its development of smart communities to its national territory but has also been active abroad. This internationalization of Japanese urban smart experiments was mainly driven by large groups supported by government agencies such as NEDO (New Energy and Industrial Technology Development Organization, equivalent to ADEME in France, the French Ecological Transition Agency). NEDO has thus supported the development of six Japanese experimentations abroad in Hawaii, Java, Los Alamos (USA), Manchester, Malaga and Lyon. These projects are based on heavy investments from Japanese companies: Toshiba, Mitsubishi Electric and Mitsubishi Motors. As such, Toshiba is carrying out 35 projects abroad, including the Lyon smart community project, which is an experiment included in the larger Lyon Confluence project – the largest city center requalification project in continental Europe.

The internationalization of Japanese smart community experiments has dual potentiality: experimentations abroad that offer Japan feedback on experiences and the export of a mature industrial solution and technological know-how in a pure market logic. For the moment, we are mainly observing the first option. Multiplying initiatives abroad is indeed a major challenge for Japanese companies during the technology and industrial product development phases. There are several reasons for this. Future access to foreign markets, including the French market, is an important point that is currently being prepared, with the establishment of cooperation with French companies in the energy sector, such as EDF, Veolia and Engie, as well as such other sectors as construction (with VINCI or Bouygues) or telecommunications (Orange or SFR). Seeking co-financing of pilot projects from partner companies and foreign governments also helps to reduce the costs of research and technological innovation. The cross-fertilization of advanced Japanese and French technologies is also an important element. The study of the reaction of test markets, particularly the reception and use of new technologies in various socio-cultural contexts, allows for the anticipation of future rejections.

Finally, investing in test projects abroad helps to develop interoperability between technologies and impose Japanese companies' own standards on their competitors. The communication protocols among the different elements of the "smart city", which have not yet been finalized, represent a point of contention between Japanese firms and a key point in the experiments in progress. In so-called "smart" homes, the "Echonet Lite" standard for communication between home electronic devices, promoted by Toshiba, was finally adopted, with the support of METI, and had to be taken up by Japanese companies, which neither had it nor had, so far, adopted it, like Panasonic.[5] From the point of view of communication between buildings, the protocol adopted by Toshiba, as well as by several American firms, is called "Open ADR", and the company is promoting it within the international organization in charge of standardization, the International Electronic Commission (CEI or IEC), and through the international network Open ADR Alliance (Granier, 2018).

To date, Toshiba has not yet developed a fully "smart" city. The company tests partial aspects in pilot spaces that, end to end, explore the different aspects of a hypothetical smart city or smart community. Two projects have been particularly successful and important for the company: Yokohama in Japan and Lyon Confluence in France, the technological form of which is almost identical in both projects – a community energy management system (CEMS) connected to buildings and new generation transport systems. The development of demonstrations, such as Yokohama and Lyon, must therefore be included in this logic of verifying technologies with a view to their international standardization.

Community energy management systems are articulated at several scales: that of the community with the CEMS as a whole and at the scale of the building (BEMS), the house (HEMS) and the factory (FEMS), generally equipped with a consumption display device (e.g., tablet, wall display). Each of these energy management systems is connected to CEMS, which centralizes all data and forecasts consumption peaks based on consumption profiles and weather forecasts. The CEMS then sends out a request to households or office workers to reduce their consumption during peak periods in summer and winter[6] (Leprêtre, 2018).

In Lyon, the smart community project was carried out from 2012 to 2017. It consisted of several experiments, including the construction of an intelligent neighborhood called Hikari (which means "light" in Japanese, both in reference to the French-Japanese partnership, but also speaking to the theme of electricity and the intelligent micro-grid as well as, more philosophically, the idea of knowledge and discovery). Added to this were Sunmoov', an innovative transport project comprised of a fleet of 30 car-sharing electric vehicles, Consotab for the visualization of the energy consumption of homes and buildings following the energy refurbishment of the old real estate stock and a community management system for data management.

The Lyon smart community program is a French-Japanese initiative, which brought together the Métropole de Lyon and the NEDO, around 30 French and

Japanese partners, including the SPL Lyon Confluence and the Toshiba group for the general coordination of the project and the implementation of technologies, Bouygues Immobilier for the construction of Hikari, Transdev for Sunmoov' and Grand Lyon Habitat for Consotab. Energy was at the heart of the experiments. The Hikari block, consisting of a set of three mixed housing, retail and office buildings, is efficient from an energy point of view. In addition to photovoltaic panels, the set has a heating plant running on vegetable oil and a cold production system, thanks to the absorption of water from the Saône. A smart grid also makes it possible to switch energy production and consumption according to the time of day and needs of buildings' users. Ultimately, Hikari buildings consumed 50% to 60% less energy than a conventional building. In addition, 67% of electrical needs, 87% of heating needs and 80% of air conditioning needs were self-produced.

The existing housing stock that has been eco-renovated has also been fitted with the Consotab system. Composed of digital tablets, it allows residents to monitor their energy consumption in real time. Residents were also made aware of good practices to reduce energy consumption, which led to a reduction in energy consumption of 8% in summer and up to 12% in winter. Finally, a community management system (CMS) has been deployed throughout the Lyon smart community: this is a global management and monitoring system for energy data, which enables energy data from Hikari, Sunmoov' and Consotab to be centralized, visualized and analyzed. The data is aggregated with other data collected by the Métropole de Lyon, such as weather and air quality, to produce dashboards used by the community, as well as by individuals, to make energy decisions. The system is therefore virtually identical to the one implemented in Japan in its own smart communities, especially in Yokohama and Kashiwa no Ha.

Japanese smart communities, a model for French cities' energy planning?

Beyond punctual collaborations between France and Japan, the question arises as to whether Japanese smart communities can serve as a model for the development of French smart cities. The latter are indeed, for the moment, very disparate. The Lyon project is quite unique in France, and compared with the great French disparity of smart cities, Japanese smart communities seem to be the result of a more uniform and better-controlled methodology.

Contrary to what can be observed in North America and Europe, the role of the state in the establishment of Japanese smart communities has been essential (Leprêtre, 2018). The central government is not only at the origin of the formulation of industrial policies and generic strategies, issues and objectives at the national level, but also responsible for the wide dissemination of smart communities abroad. Japanese government programs, relayed in most cases by such national agencies as NEDO, are then invested in by consortia of private actors, led by national champions of digital technologies, in partnership with local public actors (municipalities, more rarely departments in Japan). In other words, the rapid multiplication of smart community experiments is the result of a national industrial strategy and privileged relations between the government and its national champions under a developmentalist capitalist regime.

Japan's "smart revolution" is the product of four factors that deeply mark the way in which smart communities are structured: (1) a national political program defined by the state and an outgoing industrial strategy; (2) a long-term entrenchment of industrial ICT policies since at least the 1970s; (3) the need to find new outlets for industrial investments to exceed the glass ceiling for the growth of mature sectors; and (4) the Fukushima disaster, which vigorously directed the national effort toward energy issues and the question of transition. Can these specificities lead, for other countries, to the formalization of a model that is ultimately quite specifically Japanese? Only a cultural anthropology of the socio-technical devices of French smart cities and Japanese smart communities can answer this question.

Conclusion: for a multi-scalar governance and a life-cycle approach of French smart cities

Smart cities and smart communities have multiplied in France and Japan, most of the time in the form of experimental urban projects. This urban planning by experimentation has led to four highly problematic issues: a fragmentation of the urban into projects that, most of the time, escape public debate; privatization of the development and governance of urban information systems; popular participation in the production of technologies through active and passive consumption practices; and the transformation of the urban into resources that can be mobilized in experimentation.

Energy policies inherited from French urban sustainability led to French smart city initiatives. However, French smart cities are not sustainable cities augmented by digital technology. Rather, they result from political issues and a powerful generational succession in the field of stakeholders mobilized in the urban fabric. From a political point of view, as well as from a technological point of view, one of the main energy issues of smart cities is the articulation of autonomous energy micro-grids with regional and national technical macro-systems of energy production and distribution, particularly electricity. In this sense, the energy governance of cities requires a perspective that is no longer urban, but rather multi-scalar, articulating different levels – from the most local to the national or even continental levels, in the case of France (due to the European integration of electricity networks).

One thing is certain: smart cities are the result of more complex governance and technological characteristics of urban energy networks and prepare the deregulation of energy markets in France, which Japan continues to observe attentively. Within this growing complexity, electricity is both the alpha and omega of French smart cities: it is both a target and a factor of dependence – even fragility. In this sense, the sustainability of smart cities and the achievement of objectives 7 and 11 of the SDGs must consider two realities.

First, it is important to pay particular attention to the relationship between primary and secondary energy in the construction of smart energy equipment and networks: a life-cycle approach to the digital components constituting smart cities is therefore necessary. On the other hand, it is important to establish a multi-scalar and cross-sector energy governance that takes account of this life-cycle approach of components, but also considers the involvement of a growing variety of actors in energy planning and in the fabric of the city – including foreign players, such as large Japanese companies. In this sense, only a cultural anthropology of digital technologies, energy consumption practices and ways of making cities will lead to the local realization of objectives 7 and 11 of the SDGs in France.

Notes

1 On this subject, see the report by La Fabrique de la Cité published in 2020: *Smart cities, débats singuliers pour un modèle pluriel. Cahier 1 – Des acteurs, des approches et des smart cities*, https://www.lafabriquedelacite.com/publications/des-acteurs-des -approches-et-des-smart-cities/
2 Proposed by Chalmers Johnson in 1982, the notion of a developmental state refers to strong links between the administration, private actors and politicians in the development and conduct of public policies promoting technological innovations.
3 There were three other programs: the "Next Generation Energy Technology Demonstration Projects" were implemented in eight cities in 2011–2012; the "Projects Supporting the Wider Acceptance of Smart Community Concepts" program consists of feasibility studies; and "smart communities" were implemented in eight towns and villages in the Tohoku region devastated by the 2011 tsunami.
4 The decentralized model is not the only component that works in favor of a post-Fukushima city model. The compactness and diversity of the functions promoted in the project also come under the same issue. Indeed, on 11 March 2011, the first blackout to have occurred was not that of energy, but that of mass transportation, with the immediate shutdown of all metro lines, leaving millions of commuters in stations. This fear of the paralysis of a sprawling city has come up again and again. The ideal of the compact city responds to this collective fear that plagues the minds of the Japanese. Kashiwa no Ha is therefore also an experimental pilot model of a post-risk city, in which the vulnerability of complex systems has been drastically reduced (at least in the minds of its promoters).
5 Panasonic is Toshiba's direct competitor for setting technical standards and accessing the "smart" market. Historically, the greater Tokyo region (Kantô) is an industrial and economic stronghold of Toshiba, while Panasonic operates in Kansai (Osaka-Kyoto-Kobe conurbation). However, Panasonic recently seized the opportunity of "smart cities" to penetrate the stronghold of Toshiba by developing a "smart city" (Fujisawa Sustainable Smart Town) and by entering the "smart" market in Yokohama. Among the 4,000 HEMS put in place, Panasonic managed to provide a greater number of HEMS to the voluntary participants than Toshiba, its partner and competitor in this project.
6 In Japan, these requests essentially take the form of dynamic pricing, with an actual price change in Kitakyushu (from 15 y/kWh to 150 y/kWh depending on the time of use and the strength of the peak) and below the form of a point system in the other three smart communities (with a loss or gain of points, returned in the form of money at the end of the season). See Leprêtre (2018) for more details.

References

Baraud-Serfaty, I., in Damon, J., Denis, E. and Strauch, L. (2013) *Smart cities: efficaces, innovantes, participatives, comment rendre la ville plus intelligente?* Report, Institut de l'Entreprise, p. 51.

Batty, M. (2013) 'Big data, smart cities and city planning', *Dialogues in Human Geography*, 3–3, pp. 274–279.

Courmont, A. and Le Galès, P. (2019) *Gouverner la ville numérique*. Paris: PUF, 120 p.

Dupuy, G. (2014) 'L'avenir de la smart city', *Urbanisme*, 394, pp. 34–35.

Erlinghagen, S. and Markard, J. (2012) 'Smart grids and the transformation of the electricity sector: ICT firms as potential catalysts for sectoral change', *Energy Policy*, 51, pp. 895–906.

Eveno, E. (1997) *Les pouvoirs urbains face aux technologies d'information et de communication*. Paris: PUF, 127 p.

Eveno, E. (2014) 'Comment l'intelligence vînt aux villes', *Urbanisme*, 394, pp. 26–27.

Gram-Hanssen, K. (2008) 'Consuming technologies – developing routines', *Journal of Cleaner Production*, 16, pp. 1181–1189.

Granier, B. (2018) 'Smart cities et gouvernementalisation de la consommation d'énergie domestique au Japon. Le rôle central de l'accident de Fukushima et des pratiques étasuniennes', *Flux*, 114(4), pp. 56–70.

Languillon-Aussel, R. (2018) 'Le programme « smart communities » au Japon. Nouveaux enjeux de pouvoir des ressources et des systèmes d'information urbains', *Flux*, 114(4), pp. 38–55.

Languillon-Aussel, R., Lepretre, N. and Granier, N. (2016) 'La stratégie de la "smart city" au Japon: expérimentations nationales et circulations globales', *Echogéo*, 36. https://echogeo.revues.org/14598 (last access in July 2020).

Leprêtre, N. (2018) 'Un « modèle national » de ville intelligente? Le rôle de l'État dans la mise en œuvre de réseaux électriques intelligents au Japon', *Flux*, 114(4), pp. 9–21.

Picon, A. (2015) *Smart Cities: A Spatialised Intelligence*. New-York: Wiley, 168 p.

Shove, E. (2010) 'Beyond the ABC: Climate change policy and theories of social change', *Environment and Planning*, 42, pp. 1273–1285.

United Nations, Department of Economic and Social Affairs, Population Division (2009) *World population prospects: The 2008 revision, highlights*, 107 p. https://grist.files.wordpress.com/2009/06/wpp2008_highlights.pdf (last access in July 2020).

12 Analysis of supply–demand balances in western Japan grids in 2030

Integrating large-scale photovoltaic and wind energies: challenges in cross-regional interconnections

Asami Takehama and Manabu Utagawa

Introduction

This chapter aims to evaluate the feasibility of a 40% share of electricity from renewable sources in 2030 in western Japan grids and to assess the impact of high penetration of variable renewable energies (VREs) on the supply–demand balance in power supply grids. This study uses a simplified model of Unit Commitment with Economic Load Dispatching (UC-ELD) for thermal power plants (Kato, Kawai, Suzuoki, 2013; Komiyama and Fujii, 2017; Ogimoto, et al. 2017). This model assesses the supply–demand grid balance with the integration of large-scale photovoltaic (PV) and wind energies, with zero nuclear power in operation, and the effects of utilizing cross-regional interconnection lines (hereafter, referred to as "interconnections").

To help with readers' understanding, the chapter is organized into two parts. The first part presents the regulatory framework, that is, the background information concerning the regulatory framework for promoting renewable energy sources and the rules for power supply network systems. It also provides a brief explanation of the current schemes concerning grid connection of renewable generators, the transmission of renewable electricity and capacity expansions of power supply networks, and highlights barriers to renewable generators in priority transmission and in capacity use of interconnections. The second part introduces the methods for UC-ELD and supply–demand analysis as well as simulation results of supply–demand balances and renewable electricity shares in western Japan grids.

Regulatory framework of renewable energies and grid integration: the national energy plan and historical trends of renewable energy development

Historical trends in electricity generation

Figure 12.1 shows the share of electricity generation by source in IEA countries in 2017. The renewable electricity supply in Japan is 8% of total generation,

DOI: 10.4324/9781003025962-18

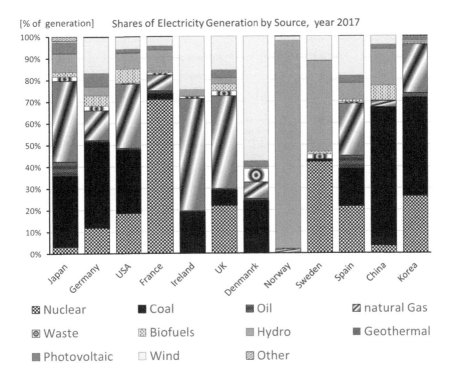

Figure 12.1 Share of electricity generation by source in IEA countries in 2017. Source: IEA data.

excluding large hydropower. The share of renewable power including large hydropower accounts for 16%. The shares of renewable generation are PV 5.2%, wind 0.6% and biomass/biofuel 2%. Japan relies on fossil fuels for around 77% of the total electricity supply. Nuclear electricity has shrunk to 3% of total generation in 2017 since the Fukushima Daiichi nuclear power plant disaster in March 2011 (hereafter, the "Fukushima nuclear disaster").

Figure 12.2 shows the Long-Term Supply-Demand Energy Outlook issued in 2015. This is still the official energy plan for the long-term electricity supply until 2030 (hereafter, "the Outlook").[1] The Outlook projects conservatively that the share of renewable electricity and large hydropower in 2030 will be 22% to 24% of total electricity generation, of which the shares of each renewable energy source are solar PV 7%, wind power 1.7%, biomass 3.7% and hydropower (including large hydropower) 8.8%. Wind power is given a miserably small share in comparison with solar PV. In contrast, the Japanese government persistently projects keeping nuclear energy at a 20% to 22% share, even in 2030.

Figure 12.3 shows the historical trends of electricity generation of the amount of energy, and Figure 12.4 shows the cumulative capacity of renewable generators before and after the feed-in tariff (FIT) scheme. Since the FIT scheme started after

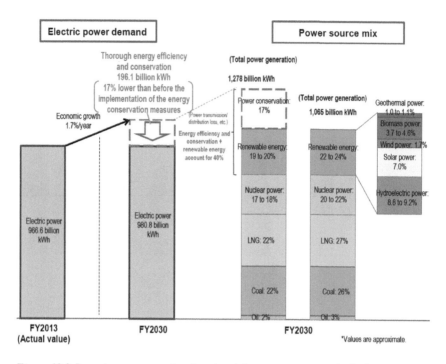

Figure 12.2 Long-term energy plan for electricity power source mix in Japan. Source: METI (Ministry of Economy Trade and Industry): Long-Term Energy Supply and Demand Outlook, 2015.

the Fukushima nuclear disaster in July 2012, solar PV energy has been increasing rapidly. However, wind energy has had a modest increase, mainly because of the many difficulties of grid access acquisition and a limited allocation of transmission capacities of 66 kV and 154 kV voltage grids, into which many wind farms need to integrate.

Positioning of nuclear and renewable energies in national energy plans before and after the Fukushima nuclear accident

This section summarizes the changes in the long-term energy plans in the past decade. Table 12.1 shows the historical changes in the positioning of nuclear and renewable energies in METI's main energy plans. Before the nuclear accident in 2011, METI's Long-Term Energy Supply-Demand Outlook in 2009 (hereafter, "the 2009 Outlook") planned to increase nuclear energy capacity from 49 GW to 61.5 GW by 2020, as "the scenario of maximum nuclear installation". The Basic Energy Plan in 2010 forecasted the construction of 14 nuclear reactors by 2030, including 9 reactors constructed by 2020. At the time, the eight power companies were planning and building the 14 nuclear reactors. On the other hand, the 2009

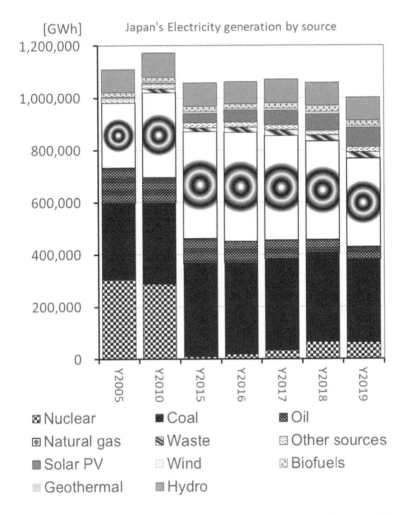

Figure 12.3 Historical trends of electricity generation in Japan by source. Source: IEA data.

Outlook projected a renewable electricity share conservatively at merely 5% by 2020 and 10% by 2030.

After the Fukushima nuclear accident, the 2015 Outlook projected a nuclear share at 22% by 2030,[2] even though the real share of nuclear electricity was 1% in 2015. On the other hand, the 2015 Outlook conservatively projected the share of renewable electricity to be 24%, of which PV is 7% and wind power merely 1.7% in 2030. The 2021 Basic Energy Plan[3] has recently been issued in July 2021. It projects renewable electricity at 22–24% and non-fossil power (nuclear + renewable) at 44% in 2030. This means the 2021 Basic Energy Plan still projects nuclear power at 20–22% in 2030. The 2021 Basic Plan has no progress in the

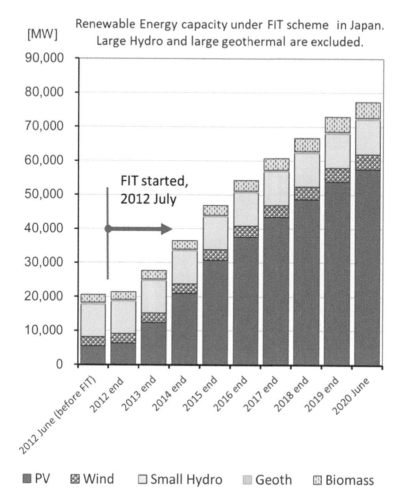

Figure 12.4 Cumulative capacity of renewable energies in Japan before and after the FIT scheme. Source: Data obtained from METI. FIT data 2012–2020. Data before FIT is based on a METI document: the committee for large-scale installation of renewable energies and next generation power grid.

renewable electricity target for 2030 from the 2015 Outlook. Japan's government still persistently seeks to keep nuclear power.

The nuclear power share at 22% in 2030 implies that the 2021 Basic Energy Plan seeks a risky way for the life extension of nuclear reactors to a longer period than 40 years and/or constructions of new reactors. This study estimates that if all of the existing nuclear reactors are to be decommissioned at 40 years old, the share of nuclear electricity in 2030 would be 10% to 12%. Without considering a life extension of nuclear reactors longer than 40 years, the 22% nuclear share

Table 12.1 The positioning of nuclear and renewable energies in the main energy plans

2009 Long-Term Energy Supply-Demand Outlook (Revised)		**2005** real	**2020** forecast	**2030** forecast
	Nuclear power capacity [GW]	49. 6	61. 5	61.5
	Share in generation [% of GWh] Nuclear	31%	44%	49%
	Renewable	1%	S%	10%

2010 Basic Energy Plan	Zero-emission power (nuclear + renewables + hydro) would supply electricity at 50% of the total electricity by 2020 New construction of the 14 of nuclear reactors by 2030 (9 reactors by 2020)

2015 Long-Term Energy Supply-Demand Outlook		**2010** real	**2030 forecast**
	Nuclear Power Capacity [GW]	48. 9	
	Share in generation [% of GWh] Nuclear	25%	22%
	Renewable	10%°	24%
			PV 7%, Wind 1.7%, Hydro 9%, Bio 4.6%, Geoth 1%
	Coal	27%	26%
	LNG, gas, oil	36%	3O%

2021 Basic Energy Plan		**2016** real	**2030 forecast**
	Share in generation [% of GWh] Non-fossil	16%	44%
	Nuclear	1%	22–24%
	Renewable	15%	

Sources: METI: long-term energy supply–demand outlook (revision) (2009). METI: the Basic Energy Plan (2010). METI: long-term energy supply–demand outlook (2015). METI: the Basic Energy Plan (2021).

would not be realistic in 2030.[4] This implies that Japan's government persistently seeks to hold nuclear energy until 2030.

Unbundling of transmission grid systems from generation businesses: the main structure of interconnections

Table 12.2 summarizes the recent regulatory framework related to renewable energies and the grid rules of the power grid system. Figure 12.5 shows the nine transmission systems operator (TSO) zones and interconnections between TSOs (note: there are ten TSOs in Japan, including Okinawa). The southern and western regions have rich potential for solar energy and the northern regions have great potential for wind energy. Figure 12.6 shows the main structure of the interconnections between the nine TSO grids. The voltage levels of interconnections and main transmission lines are mostly at 500 kV. The eastern grids (Tokyo, Tohoku and Hokkaido) are based at 50 Hz as the standard frequency, and the western grids (Chubu, Kansai,

Table 12.2 Regulatory framework of renewable energies and grid rules for power grid systems

Support scheme for RE electricity	FIT (feed-in tariffs) started July 2012 Fixed tariffs for 20 years for PV, wind, biomass and small hydro; 15 years for geothermal; ten years for small PV
TSOs and control areas	Nine TSOs, nine control zones (ten TSOs including Okinawa)
Unbundling of TSO (separation of transmission grids from generation businesses)	Unbundling came into force in April 2020. Legal separation
	Before unbundling, there were nine major generation companies (Hokkaido Electric Power, Tohoku Electric Power, Tokyo Electric Power, Hokuriku Electric Power, Chubu Electric Power, Kansai Electric Power, Chugoku Electric Power, Shikoku Electric Power, Kyushu Electric Power)
TSO independency	Before April 2020: The nine major generation companies operated transmission grid systems. They were "vertically integrated utility companies" and operated from generation, transmission and distribution sectors to the retail sector Currently, the nine TSOs are 100% subsidiaries of the nine major generation companies
Grid connection of RE generators	FIT law obligates TSOs for RE connections TSOs can reject or suspend RE connections when grid capacity is limited Voltage levels of grid integration: PV at 0.2 kV, 6.6 kV, 22 kV and 33 kV Wind energy at 66 kV FIT law does not obligate TSOs to expand grid capacity
Grid capacity development/ expansion	FIT law does not obligate TSOs to increase grid capacity to accommodate RE generators
Cost allocation of grid expansion	RE operators are required to pay a major part of grid development costs, including expansion of upper voltage grids
Priority transmission of RE power to upper voltage grids	RE power has no priority to be transported in transmission lines (500 kV and 275 kV) and interconnections FIT law has no definition of transmission of RE electricity with priority at the first and second voltage level grids and to neighboring TSO zones At the highest-voltage grids (500 kV) and the second-highest voltage grids (275 kV), "first-come generators" have a priority in utilizing transmission capacity
Cross-regional interconnections	Cross-regional interconnections between nine TSO grids

(Continued)

Table 12.2 Continued

Priority transmission of RE power through interconnections	FIT law has no definition Before October 2018, "long-term fixed generators" (= nuclear power), "scheduled power supply" (= coal power), and "first-come generators" had a priority in interconnection use Since October 2018, interconnection capability is used according to the merit order of electricity trade prices ("connect and manage" rule) Nuclear power substantially has a priority in interconnection use as an exception, even now RE power has no priority in interconnection use
Curtailment, reduction of RE power feed- in (feed-in management)	FIT law allows TSOs to reduce feed-in power from RE generators when RE power supply exceeds the demand in each TSO control zone RE operators got little compensation for curtailment, being compensated only when curtailment continues for more than 30 days. For difficulties in supply–demand grid balance, there is no compensation
Nuclear generation businesses	The nine major generation companies own almost all of the nuclear generators

Source: Data and documents from METI.

Hokuriku, Chugoku, Shikoku and Kyushu) are based at 60 Hz, with frequency convertors at the interconnections between the Tokyo and Chubu zones.

Unbundling of transmission grids, which means a separation of transmission grid systems from power generation businesses, is one of the key factors facilitating grid integration of renewable energies. The unbundling of transmission and distribution grids came into force just recently, in April 2020, as a legal separation.[5] Before April 2020, the nine major generation companies owned and operated transmission and distribution grids. Before the unbundling, the nine major generation companies were "vertically integrated power suppliers" that integrated power generation, transmission, distribution and retail businesses. Even now, these companies and their subsidiaries still own or operate 99% of nuclear power capacity, 66% of coal power capacity and 93% of liquefied natural gas (LNG) thermal power capacity [% of MW] in relation to the total generation capacity of generation businesses as of March 2019.[6] Currently, the nine TSOs are 100% subsidiaries of the nine major generation companies. These facts imply insufficient independence of TSOs from the major generation businesses.

Absence of grid expansion and cost of grid development for renewable businesses

The FIT scheme started in July 2012.[7] This scheme is based on the "Act on Special Measures Concerning Procurement of Electricity from Renewable

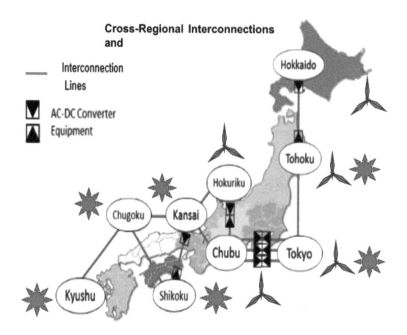

Figure 12.5 The nine TSO control zones and geographical patterns of wind and solar energy potentials. Source: TSO zone map from the Organization for Cross-Regional Coordination of Transmission Operators (OCCTO), "Outlook of Electricity Supply-Demand and Cross-Regional Interconnection Lines" (2019). Solar and wind figures simplify geographical patterns of renewable potentials in each zone. Solar and wind figures are attached for easy understanding.

Energy Sources by Electricity Utilities"[8] (hereafter, the "FIT law"). The FIT scheme gives fixed tariffs for 20 years for electricity from PV, wind, biomass and small-hydro systems, 15 years for geothermal and ten years for PV systems smaller than 10 kW.

Concerning the grid connections of renewable generators, the FIT law requires TSOs to connect renewable generators as an obligation. However, the FIT law does not require TSOs to expand grid capacity. Because of this loophole, the FIT law allows TSOs to reject or suspend grid connections of renewable generators when line capacity is not sufficient in transmission and distribution areas.

Table 12.3 shows examples of grid voltage levels and renewable energy integration in TSO zones. Because the FIT law does not require TSOs to increase grid capacity to accommodate renewable generators, renewable energy developers bear expenses for a major portion of grid development costs, including expansion costs of upper voltage grids (such as additional transformers, additional lines in the

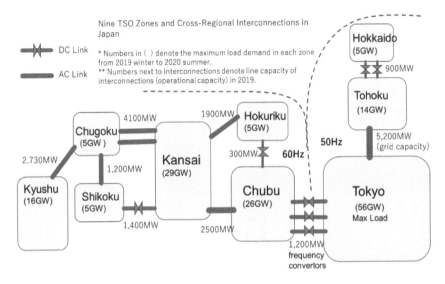

Figure 12.6 The main structure of cross-regional interconnections between nine TSO grids. Source: Documents from the Ministry of Economy, Trade and Industry (METI): "The next generation grid network committee". Numbers in ( ) denote the maximum load demand in each TSO zone in winter 2019 and summer 2020. Numbers next to interconnections denote line capacity (aggregated operational capacity) in August 2019.

existing grids and those in the upper voltage grids).[9] This expense of grid expansion is allocated to renewable developers, in addition to the expenses of a power source line from renewable generators, according to the Ministry of Economy, Trade and Industry (METI) guidelines for the cost allocation of grid development.[10]

This rule causes serious difficulties to renewable energy developers, especially wind energy developers. Generally speaking, the cost of grid development in 66 kV and 154 kV grids for integrating wind farms is much greater than the cost in 6.6 kV grids for integrating PV systems. This situation has led to a significant stagnation in grid development integrating renewable energies, and especially integrating wind energies, even though wind energy has guaranteed fixed tariffs under the FIT scheme. Before the unbundling in 2020, the nine major generation companies commanded transmission grid systems and they restricted the grid connection of wind farms with the excuse of a shortage of grid capacity.

Large-scale conventional power plants (nuclear, coal and LNG thermal plants) are connected to 500 kV and 275 kV grids. The grid development costs for these voltage levels (the highest and second-highest voltage grids) are covered by grid charges from electricity consumers. This is the definition by the METI guidelines.[11]

Table 12.3 Grid voltage levels and renewable energy integrations in TSO zones

	Hokkaido	Tohoku	Tokyo	Chubu	Kansai	Kyushu	Cross-regional interconnections (OCCTO)	Voltage levels integrating renewable generators
Main transmission	187 kV	500 kV 275 kV	500 kV 275 kV	500 kV 275 kV	500 kV 275 kV	500 kV 220 kV	500 kV 275 kV	
Sub-transmission	110 kV 66 kV 33 kV, 22 kV	154 kV 66 kV 33 kV	154 kV 66 kV 22 kV	154 kV 77 kV 33 kV, 22 kV	154 kV 77 kV 33 kV, 22 kV	110 kV 66 kV 22 kV		Wind Wind, PV (large) Wind, PV
Distribution			6.6 kV 200 V, 100 V			6 kV		PV PV (residential)

Source: This is summarized of data from TSOs.

Renewable electricity without priority in transmission grids and interconnections

Rules for capacity use in interconnections

Renewable power has no priority to be transported through transmission lines and interconnections. The FIT law does not define transmission and upstream transmission of renewable power with a priority to the highest and second-highest voltage transmission grids (500 kV, 275 kV) or to neighboring TSO grids through interconnections.[12]

Concerning the grid rules (grid operational rules) for interconnection use, until October 2018 the allocation of transmission capability of interconnections was based on a "first-come, first-served basis".[13] This meant that "the first-come" generators had priority in interconnection transmission. Large-scale conventional power plants (mainly coal and nuclear power plants) reserved a major portion of interconnection capacity one year or ten years ahead. Little capability was left for "late-come" renewable generators in interconnection use. "Long-term fixed generators", which were mainly nuclear generators, and "scheduled power supply" mainly from coal and nuclear plants had priority in interconnection use.

A new grid rule, the "Implicit Auction Scheme" for interconnection use, started in October 2018. Under this rule, the whole capacity of interconnections is traded in the day-ahead spot market (Japan Electric Power Exchange [JEPX] Spot Market). Renewable generators can use the capacity of interconnections in accordance with the day-ahead spot price. However, even under this new rule, renewable generators have no priority in utilizing interconnections. In contrast, nuclear power plants are still guaranteed to utilize the capacity of interconnections as an "exception".[14]

Rules for capacity use in transmission grid inside each TSO zone

Concerning the grid rules for the highest (500 kV) and second-highest (275 kV or 200 kV) voltage grids inside each TSO zone, "first-come generators" still have priority in the utilization of transmission capacity.[15] This leads to a limited amount of remaining transmission capacity for renewable generators. When capacity in transmission lines is congested, curtailment (output reduction) is imposed on renewable generators because these are "late-come" generators.[16]

Curtailment (output reduction) of renewable electricity

The FIT law states that TSOs can impose curtailment (output reduction) of feed-in power from renewable generators when renewable electricity supply exceeds "an amount of demand in *each TSO control zone*".[17] This implies that nationwide transmission of renewable electricity through interconnections with a priority is not guaranteed under the FIT law and related regulations. In addition, renewable operators are given zero or very little compensation for curtailment up to 30 days, or even for more than 30 days, depending on conditions.[18,19]

Under the FIT law, at a time of curtailment of renewable electricity, fossil fuel power is inferior to renewable electricity and fossil power is reduced before renewable electricity is reduced in each TSO zone. However, nuclear power always has priority to feed-in, even at a time of oversupply in a TSO zone.[20]

The above aspects of the FIT law and the grid rules imply that priority for renewable electricity is implemented in limited situations. (a) Priority feed-in is implemented only when it is against fossil fuel electricity. (b) Priority transmission is implemented only when it is inside each TSO zone. (c) Renewable electricity transmission is not secured at the highest and second-highest voltage transmission lines (500 kV and 275 kV) in each TSO zone and in interconnections. All these situations have had serious impacts on the stagnation of wind energy development, even with FIT tariffs.

Grid development plan of interconnections

Table 12.4 shows the operational capacity of interconnections in 2020 and the planned capacity (in thermal capacity) in 2029 in OCCTO's long-term grid development plan. Capacity development of interconnections is one of the key

Table 12.4 Grid capacity of cross-regional interconnections in 2020 and COOTO's grid development plan for 2029

Direction			[MW]	
			Y2020 Aug	Y2029, OCCTO's planned capacity
Hokkaido	>>	Tohoku	900	900
Tohoku	>>	Tokyo	6,150	10,280
Tokyo	<< >>	Chubu	1,200	3,000
Kansai	>>	Chubu	2,500	2,500
Chubu	>>	Kansai	1,240	1,240
Kansai	>>	Chugoku	2,500	2,500
Hokuriku	>>	Kansai	1,900	1,900
Chugoku	>>	Kansai	4,210	4,210
Shikoku	>>	Kansai	1,400	1,400
Shikoku	>>	Chugoku	1,200	1,200
Kyushu	>>	Chugoku	2,380	2,780

Interconnection operational capacity can vary depending on seasons, daytime or nighttime, holiday or business day.

Source: Data from OCCTO, the Committee for Cross-Regional Interconnections Network Development Plan: document on "The operational capacity of interconnections (long-term plan) for 2020 to 2029".

factors in increasing grid integration of renewable energy systems. The long-term plan shows that only the interconnection capacity of Tohoku to Tokyo will be developed significantly. The capacity increase for Kyushu to Chugoku is on a minor scale. The capacity increase for Hokkaido to Tohoku has yet to be decided, although the PV potential is rich in the Kyushu, Chugoku and Shikoku zones and wind potential is very rich in the Hokkaido and Tohoku zones.

OCCTO's long-term plan shows a zero increase in grid capacity for Hokkaido to Tohoku and Shikoku to Kansai, and a limited scale of capacity increase for Kyushu to Chugoku. These plans imply that OCCTO and the nine TSOs have a limited interest in transporting renewable electricity nationwide.

Methods

Points for analysis

As described earlier, inter-regional transmission with priority is not guaranteed for renewable electricity under the FIT law and the OCCTO's grid rules. Contrary to the current framework, the UC-ELD model described in this section simulates a future scenario where the first priority of transmission is sufficiently guaranteed for renewable electricity in interconnections, and where a large capacity of PV and wind energies are integrated into the grids. The model estimates the following items:

- Whether is it possible to achieve a renewable electricity share of 40% per generation in 2030?[21]
- The amount of power oversupply or power shortage that occurs with a large capacity of PV and wind energies and zero nuclear power in operation.
- The amount of electricity that can be transmitted from the Kyushu, Shikoku and Chugoku zones to the Kansai-Chubu zone where a large-scale electricity demand exists, if interconnection capacity is fully utilized to transport renewable electricity.

The UC-ELD model and points of analysis

Using a simplified UC-ELD model, conventional generator units in each control zone were classified into 22 subgroups of generators. This classification depended on the techno-economical parameters of the generation units, such as load-following capability in increasing or decreasing output, lower output limit and availability for load-frequency control power (LFC) of generation units.

The main generator subgroups are coal 1, 2, 3; oil 1, 2, 3; LNG thermal 1, 2, 3; gas-CC 1, 2, 3; independent power producers 1, 2, 3; pumped storage hydropower (PSHP: capacity aggregated into one virtual generator); hydro-reservoir (capacity aggregated into one generator); and nuclear (capacity aggregated into one generator). The UC-ELD model uses aggregated values in capacity of each subgroup in [MW].

The UC-ELD model was used to estimate the hourly power supply–demand balance in May (a low demand period) and August (a high demand period) for 2030 in the five western control zones (Kyushu, Shikoku, Chugoku, Kansai and Chubu). Actual grid data in 2016 was used as a base year. The simulation does not include the Hokuriku zone due to the limited accuracy of wind power forecasts in our model.

Formulation

Exogenous variables

k: power plant subgroup
t: time [hour]
C_k: rated capacity of conventional power plant subgroup k (aggregated capacity) [MW]
u_k: binary variable indicating on or off plant subgroup k at time t. $u_k = 0$ or 1
$PHy(t)$: power output from run of river hydro [MW]
$D(t)$: hourly electricity demand at time t [MW]
$PPV(t)$: PV power output at time t [MW]
$PWd(t)$: wind power output at time t [MW]
$PBioGe(t)$: biomass and geothermal power output at time t [MW]
$REIm(t)$ and $REEx(t)$: renewable power transmission through interconnections (import and export) at time t
$Pmin_k$: minimum real power output of subgroup k
$Pmax_k$: maximum real power output of subgroup k
FK_k: fuel cost of plant subgroup k [JPY/kWh]
m: dummy price of power supply shortage
UR_k or DR_k: maximum up ramp or down ramp rate of plant subgroup k [pu/h]
w_k: own consumption ratio of plant subgroup k
$LHP(t)$: loading of heat pump systems (HPs) at time t
$LEV(t)$: charging electric vehicles (EVs) at time t
$CRps_k$ or $CRng_k$: ratio of positive or negative LFC control reserve relative to rated capacity of plant subgroup k

Endogenous variables

$P_k(t)$: hourly power output of plant subgroup k at time t [MW]
$TK(t)$: total fuel cost at time t [JPY/h]
$S(t)$: hourly power supply shortage at time t [MW]

Constraints

Objective function: The objective function to be minimized is hourly fuel cost. The calculation was performed using Matlab Optimization Tool Box.

$$min.TK(t) = \sum_{k=1}^{N} P_k(t) \times u_k(t) \times FK_k + m \times S(t) \left[JPY / h \right] \tag{12.1}$$

Power supply–demand balancing

$$0.95 \sum_{k=1}^{N} P_k(t) \times u_k(t) \times (1 - w_k)$$

$$+ 0.95 \{ PHy(t) + PPV(t) + PWd(t) + PBioGe(t) \} + S(t) + REIm(t) \tag{12.2}$$

$$= D(t) + LHP(t) + LEV(t) + REEx(t)$$

where a total 5% loss for transmission and distribution is used for conventional plants and renewable generators. Due to the transmission of renewable electricity into the entire western grids, a 5% loss is also used for renewable generators.

LFC control reserve capacity constraints

$$\sum_{k=1}^{N} C_k \times u_k(t) \times CRps_k \geq 0.03_D(t) \tag{12.3}$$

$$\sum_{k=1}^{N} C_k \times u_k(t) \times CRng_k \geq 0.03D(t) \tag{12.4}$$

where the control reserve capacity at every hourly time point must be larger than 3% of the demand in each control zone.

$$0.05 \leq CRps_k \tag{12.5}$$

$$0.05 \leq CRng_k \tag{12.6}$$

$$C_k \times (P\min_k + CRng_k) \leq P(t) \leq C_k \times (P\max_k - CRps_k) \tag{12.7}$$

where the available LFC control reserve capacity must be larger than at least 5% of each generating subgroup. Therefore, the lower and upper output limits of each subgroup are reduced by at least 5% to keep the available power output range for the LFC control reserve. Coal-fired generation units are not utilized for the

LFC control reserve. The minimum power output limits of generating units are assumed to be 15% to 30% relative to the rated capacity for coal-fired units, 20% to 30% for LNG thermal and gas-combined cycle and 15% to 25% for oil power units. The minimum power output limits of thermal power units were defined using data from METI.[22]

Ramp up/down rate constraints

$$P_k(t) - P_k(t-1) \leq UR_k, \text{ if generation increases} \tag{12.8}$$

$$P_k(t-1) - P_k(t) \leq DR_k, \text{ if generation decreases} \tag{12.9}$$

Hourly up-ramps and down-ramps of generator subgroups are constrained in ranges: oil power subgroups 1 to 8% relative to the aggregated rated capacity per minute ([%/min]); coal power subgroups 1 to 3%; LNG thermal subgroups 3 to 8%; gas-CC (GTCC: gas turbine combined cycle) subgroups 5 to 10%. The hourly up-ramps and down-ramps of power through interconnections are limited to less than 15% of the operational capacity (transmission capability) of the interconnections. The operational capacities of interconnections are based on data from OCCTO.[23]

Simplified PV forecast and wind power forecast models are used to estimate PV and wind power outputs one-hour ahead. The model assumes that day-ahead forecast errors of PV and wind power outputs are sufficiently renewed by the time of gate-closure of intraday power trades. The gate-closure of the intraday market is one-hour ahead of the real-time power supply. Therefore, the UC-ELD model assumes that the LFC control reserve compensates for the remaining part of errors after the gate-closure; that is, one-hour ahead forecast errors of PV and wind power outputs. Demand forecast errors in one-hour ahead are assumed to be zero in this model.

Target capacity in 2030 and assumptions for a high renewable scenario

Table 12.5 shows the target capacity of the high case (Saitou, Urabe, Ogimoto, 2016) and Table 12.6 summarizes the assumptions for the Kyushu zone. The target capacity in Table 12.5 is intentionally set at a challenging level that aims to achieve 40% renewable electricity share. The high case includes the following assumptions:

All nuclear generators are assumed to be phased out in the high case. No nuclear power is used in 2030. The operational capacity of coal power is reduced as much as possible, including daily shutdowns in the daytime and in a low load period. Because coal power plants in Japan are hard-coal power generators, daily shutdown operations are possible for many hard-coal generators.

Table 12.5 Target capacity of the high case

	Demand in 2016 [GW]		2030 Targets [MW]				EV vehicles in 2030
	Max	*Min*	*PV*	*Wind*	*Heat pump*	*EV*	*[1,000 cars]*
Chubu	25	9	17,400	10,400	1,350 MW* 8h	1,120 MW*8h	2,050
Kansai	27	10	13,900	3,400	1,780 MW* 8h	890 MW*8h	1,630
Chugoku	11	5	8,000	3,200	840 MW* 8h	460 MW*8h	840
Shikoku	5	2	5,000	2,600	600 MW* 8h	240 MW*8h	440
Kyushu	16	6	18,200	4,700	810 MW* 8h	780 MW*8h	1,440

Source: Data from the Japan Wind Power Association.

Table 12.6 Targets of the high case in the Kyushu zone

Kyushu zone	[MW], Base year = 2016		
	Base	Middle	High
PV capacity	6,860	13,700	18,200
Wind capacity	490	4,700	4,700
Nuclear power	1,780	0	0
Interconnection capacity	2,690	2,690	2,780
Inter-regional transmission from PV and wind power	No	Yes	Yes
Control reserve through inter-regional lines	No	Yes	Yes
Heat pump	0	810 MW* 8h	810 MW* 8 h
Electric vehicles	0	0	780 MW* 8 h
Pumped storage hydro power	Pump-up at night, daytime generation	Daytime pump-up, evening generation	Daytime pump-up, evening generation
Pumped storage capacity	2,300	2,300	2,300
Demand (max)	15,500	15,500	−10%
Demand (min)	6,400	6,400	−10%

Source: Takehama.

HPs are in loading mode except for the evening. EV batteries are mainly in charging mode in the daytime and EV batteries are not in charging mode in the evening (16:00 to 20:00). Among passenger cars in cumulative vehicle numbers in 2030, 20% are assumed to be EVs. Electricity demand in 2030 is assumed to be decreased by 10% from the base case (the 2016 level) due to the aging of the population and energy efficiency improvement.

In the high case, renewable electricity is transported through interconnections with priority from the Kyushu, Shikoku and Chugoku zones to the Kansai and Chubu zones, because the Kansai and Chubu zones have a large scale of electricity demand due to the presence of large industrial areas and big cities. The high case considers the Kansai and Chubu zones as an integrated single zone (the Kansai-Chubu zone) in order to accommodate renewable power from the Chugoku, Shikoku and Kyushu zones.

In the current LFC control reserve scheme (as of 2020), each TSO procures and activates the major portion of the LFC control reserve in each zone only, and not through interconnections. However, the high case assumes that LFC control reserve power is activated through interconnections beyond each TSO zone. The scale of control reserve capacity must be larger than the maximum scale of wind forecast error and PV forecast error at any time. The maximum forecast error of wind and solar PV power is set at 15% of wind and PV capacity in each TSO zone.

Renewable power transmission through interconnections is assumed to be up to 80% of the operational capacity of interconnections. This is determined based on LFC control reserve exchange through interconnections. Hourly increases or decreases of power transmission through interconnections are constrained in the range of 15% of the operational capacity of interconnections by taking grid stability into account.

Figure 12.7 shows the operation modes of pumped-storage hydropower systems (PSHPs). PSHPs are in pumping mode in the daytime and in generating mode in the evening for moderating up-ramps of residual load (daytime pump-up). The daytime pump-up operation is chosen depending on the day-ahead forecast for PV output at noon. Figure 12.7 also shows an example of an hourly operation program for HPs in heating mode and EVs in charging mode.

Simulation results

Grid balance in the low load period

Figures 12.8, 12.9 and 12.10 show the supply–demand balances in May in the high case in the Kyushu, Chugoku and Shikoku zones, respectively. Negative

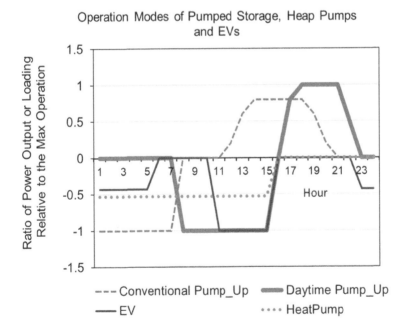

Figure 12.7 Operation modes and time schedule of daytime pump-up mode for pumped storage hydropower, EVs in charging mode and HPs in heating mode. Source: Takehama.

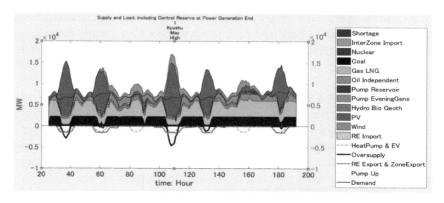

Figure 12.8 Grid balance in the Kyushu zone, 1–7 May (high case). Source: Takehama.

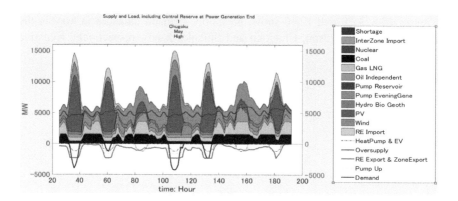

Figure 12.9 Grid balance in the Chugoku zone, 1–7 May (high case). Source: Takehama.

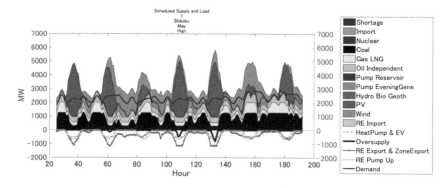

Figure 12.10 Grid balance in the Shikoku zone, 1–7 May (high case). Source: Takehama.

load in these figures indicates loads from pumping-up operations of PSHP, inter-regional transmission, HP heating and EV charging, and is shown as negative values.

Large-scale power oversupplies take place on holidays from PV power in these zones. The maximum scales of excess power supply on holidays in May are around 4 GW in the Kyushu zone and 3 GW in the Chugoku zone. In the Kyushu zone, power oversupply occurs on about half of the days in May, even with acti-vations of PSHP in daytime pump-up mode, inter-regional transmission and HPs in heating and EV charging in the daytime. Around 25% of the PV output in the Kyushu zone can be accommodated by PSHP, and 25% to 27% is transmitted to the Chugoku zone as a priority through the interconnection.

Figure 12.11 shows the supply–demand balances in May in the Kansai-Chubu zone (high case). Power oversupply in this zone occurs on a limited scale relative to its demand size, and only on holidays in the first week of May.

Utilization of inter-regional transmission in the low load period

Figure 12.12 shows that the prevailing patterns of inter-regional power transmis-sion depend on the PV output from the Kyushu to Chugoku, Chugoku to Kansai and Shikoku to Kansai zones. The largest scale of inter-regional transmission occurs from the Chugoku to Kansai zones due to the combined scales of power oversupply from the Kyushu and Chugoku zones.

Figure 12.13 shows that at a time of maximum power transmission through interconnections in May in the high case, around 2 GW is transmitted from Kyushu to Chugoku, 3 GW from Chugoku to Kansai and 1 GW from Shikoku to Kansai-Chubu. Power flow from Kyushu to Chugoku reaches the operational limit of the interconnection capability on 16 days in May. The interconnections of

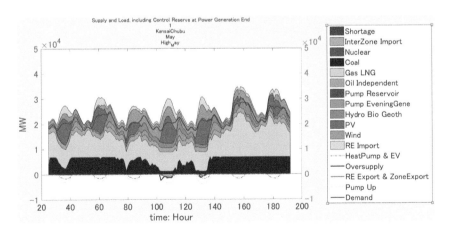

Figure 12.11 Grid balance in the Kansai-Chubu zone, 1–7 May (high case). Source: Takehama.

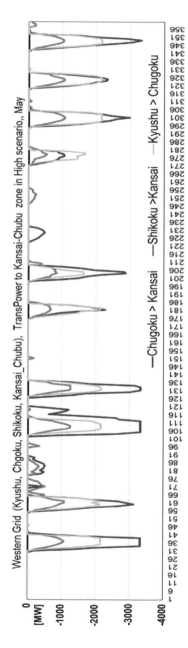

Figure 12.12 Power transmission from the Kyushu, Shikoku and Chugoku zones to the Kansai-Chubu zone, 1–14 May. Source: Takehama.

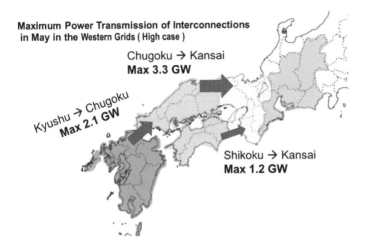

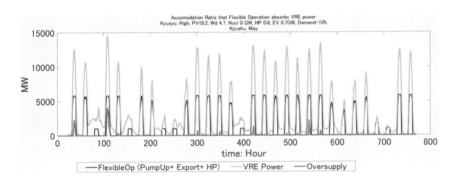

Figure 12.13 Maximum values of power transmission through inter-regional connections in May (high case). Source: Takehama.

Figure 12.14 Flexible operations absorbing VRE power and power oversupply in Kyushu in May (high case). Source: Takehama.

Kyushu to Chugoku, Chugoku to Kansai and Shikoku to Kansai would often get congested in low load periods. Capacity expansion at these three interconnections is required for transporting renewable electricity to the Kansai-Chubu zone.

To absorb power oversupply from PV output, the high case activates "flexible grid operations" that utilize PSHP, inter-regional transmission, HPs and EV charging. Figure 12.14 shows the amount of VRE power absorbed by flexible grid operations in the Kyushu zone (high case in May). From 75% to 100% of VRE power at 1:00 pm each day is absorbed through flexible operations. Power oversupply in the Kyushu zone accounts for 0% to 25% of VRE power. Flexible grid operations can accommodate the major portion of VRE power in the low load period in the Kyushu, Chugoku and Shikoku zones.

Control reserve activations in the low load period

Figure 12.15 shows LFC control reserve activations in the Kyushu zone in May in the high case. In a low load period, such as in May, the available capacity of the positive and negative control reserve power is in short supply in the Kyushu, Shikoku and Chugoku zones because the more PV capacity increases, the more PV forecast errors occur, and LFC control reserve generators need to cover an increased scale of PV forecast errors. Concerning positive control reserves in these zones, conventional generators still have space to provide a capacity of positive control reserves in May due to low levels of demand.

However, these three zones have difficulties in increasing the available capacity of the negative control reserve. An increase in negative control reserve capacity raises the minimum output levels of LFC generating units, and this reduces the capability of these units in adapting to rapid down-ramps of residual load in a time of power oversupply. Therefore, control reserve activations through interconnections are highly required and recommended.

Renewable electricity shares in the low load period

Tables 12.7 to 12.10 show renewable electricity shares, coal generation shares and CO_2 emission levels in the Kyushu, Chugoku, Shikoku and Kansai-Chubu zones, respectively, in May in the high case. The renewable electricity share in generation amount in the Chugoku and Shikoku zones exceeds the 40% level [% of MWh]. However, the renewable electricity share in the Kyushu zone is 37% due to a large amount of curtailment (output reduction) of PV electricity, and it is 23% in the Kansai-Chubu zone.

The CO_2 emission level in the high case in the Kansai-Chubu zone would increase from the base case level, although CO_2 emission in the other zones achieves a significant decrease from the base case. Renewable electricity

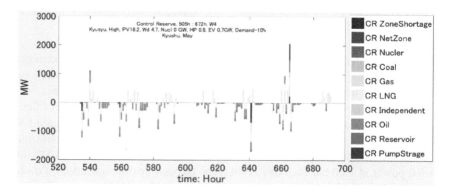

Figure 12.15 Control reserve activations in the Kyushu zone in the fourth week in May (high case). Source: Takehama.

Table 12.7 Simulation results in the Kyushu zone (May)

Kyushu zone (May)	Base	High
Renewable share in generation [% of MWh]	14.2%	36.8%
Coal share in generation [% of MWh]	34.9%	21.3%
CO_2 emissions [CO_2 kg/kWh]	0.452	0.369
Fuel cost [JPY/kWh]	7.23	7.67
Fuel cost [USD/kWh]	0.066	0.070

Table 12.8 Simulation results in the Chugoku zone (May)

Chugoku zone (May)	Base	High
Renewable share in generation [% of MWh]	12.1%	46.2%
Coal share in generation [% of MWh]	37.9%	19.4%
CO_2 emissions [CO_2 kg/kWh]	0.490	0.328
Fuel cost [JPY/kWh]	8.13	7.08
Fuel cost [USD/kWh]	0.073	0.064

Table 12.9 Simulation results in the Shikoku zone (May)

Shikoku zone (May)	Base	High
Renewable share in generation [% of MWh]	13.1%	43.9%
Coal share in generation [% of MWh]	42.6%	30.2%
CO_2 emissions [CO_2 kg/kWh]	0.437	0.379
Fuel cost [JPY/kWh]	6.122	6.412
Fuel cost [USD/kWh]	0.056	0.058

Table 12.10 Simulation results in the Kansai-Chubu zone (May)

Kansai-Chubu zone (May)	Base	High
Renewable share in generation [% of MWh]	10.8%	22.8%
Coal share in generation [% of MWh]	24.1%	23.2%
CO_2 emissions [CO_2 kg/kWh]	0.389	0.416
Fuel cost [JPY/kWh]	7.78	8.53
Fuel cost [USD/kWh]	0.071	0.078

generation in the high case in the Kansai-Chubu zone is not sufficient to offset the amount of generation from nuclear power in the base case.

Grid balances in the high load period (summer)

Figures 12.16 and 12.17 shows grid balances in the Kyushu zone and the Chugoku zone, respectively, in the high case in the first week of August, which is usually the highest load period in a year. Inter-regional electricity transmission from Kyushu to Chugoku, Shikoku to Kansai-Chubu and Chugoku to Kansai-Chubu is necessary only in the mid-day periods on a limited scale on a sunny day. Risks of supply shortage are limited in the Kyushu, Shikoku and Chugoku zones, even though zero nuclear power is in operation. This is because a sufficient amount of PV power output and energy-saving measures are available.

However, additional energy-saving measures in the evening is recommended in these zones because from 16:00 to 20:00 in August the power supply in the

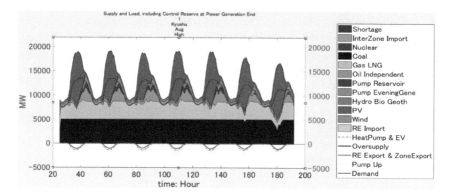

Figure 12.16 Grid balance in the Kyushu zone, 1–7 August (high case).

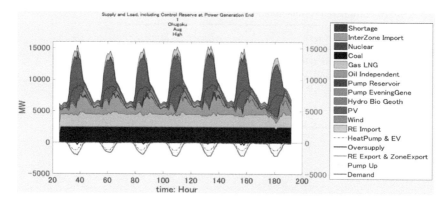

Figure 12.17 Grid balance in the Chugoku zone, 1–7 August (high case).

Kyushu and Shikoku zones becomes quite tight due to a rapid ramp-up of residual load, for which fossil fuel generators must increase output at maximum ability. In the summer in western Japan, wind energy supply is at low levels and the major portion of renewable electricity is supplied from PV power. The more PV capacity is integrated, the more rapid ramp-up of residual load occurs in the evening. This would cause a tension of supply–demand balance in a few hours in the evening. Therefore, additional energy-saving measures in evening periods are significant for securing grid balances.

$$
\text{Residual load} = \text{Electricity demand} \\
- \left(\begin{array}{l} \text{PV power} + \text{Wind power} \\ + \text{Biomass and Geothermal power} + \text{Hydropower} \end{array} \right) \quad (12.10)
$$

Figure 12.18 shows the grid balance in the Kansai-Chubu zone in August in the high case. This figure shows an additional demand decrease by 15% from the base case, rather than a 10% decrease. A reduction of demand by 10% is not sufficient for this zone in summer peak load periods. The supply capacity in the Kansai-Chubu zone becomes extremely tight in August in the evening when zero nuclear power is in operation. Risks of supply shortage would occur due to the steep ramp-up in residual load in the evening. The scales of supply shortage in August are in the range of 150 MW for a few hours, mainly in the evening. This implies that additional energy-saving measures are required in the Kansai-Chubu zone in high load periods.

Renewable electricity shares in the high load period (summer)

Tables 12.11 to 12.14 show simulation results including renewable electricity shares in August. The renewable electricity shares are 31% in Kyushu, 37% in Chugoku, 31% in Shikoku and 16% in Kansai-Chubu [% of MWh]. In the high

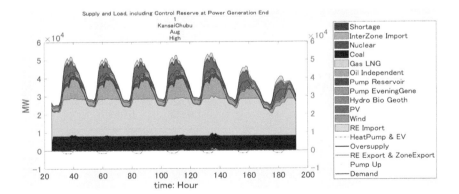

Figure 12.18 Grid balance in the Kansai-Chubu zone, 1–7 August (high case).

Table 12.11 Simulation results in the Kyushu zone (Aug)

Kyushu zone (Aug)	Base	High
Renewable share in generation [% of MWh]	12.6%	31.3%
Coal share in generation [% of MWh]	49.0%	38.3%
CO_2 emissions [CO_2 kg/kWh]	0.531	0.455
Fuel cost [JPY/kWh]	6.96	6.70
Fuel cost [USD/kWh]	0.063	0.061

Table 12.12 Simulation results in the Chugoku zone (Aug)

Chugoku zone (Aug)	Base	High
Renewable share in generation [% of MWh]	9.7%	36.5%
Coal share in generation [% of MWh]	42.8%	27.3%
CO_2 emissions [CO_2 kg/kWh]	0.553	0.406
Fuel cost [JPY/kWh]	8.78	7.51
Fuel cost [USD/kWh]	0.080	0.068

Table 12.13 Simulation results in the Shikoku zone (Aug)

Shikoku zone (Aug)	Base	High
Renewable share in generation [% of MWh]	11.3%	30.6%
Coal share in generation [% of MWh]	53.7%	52.3%
CO_2 emissions [CO_2 kg/kWh]	0.512	0.516
Fuel cost [JPY/kWh]	5.87	6.11
Fuel cost [USD/kWh]	0.054	0.056

Table 12.14 Simulation results in the Kansai-Chubu zone (Aug) with additional demand decrease by 15%

Kansai-Chubu zone (Aug)	Base	High
Renewable share in generation [% of MWh]	8.3%	15.6%
Coal share in generation [% of MWh]	20.5%	24.1%
CO_2 emissions [CO_2 kg/kWh]	0.423	0.467
Fuel cost [JPY/kWh]	9.62	9.76
Fuel cost [USD/kWh]	0.088	0.089

load period, the five western zones do not attain the 40% target. In particular, the renewable electricity share in the Kansai-Chubu zone remains far below the target level.

CO_2 emission levels in the Kansai-Chubu and Shikoku zones in the high case would increase from the base case. This is because, under the condition of zero nuclear power in operation, fossil fuel generation would increase for meeting the large scale of cooling demand during summer. To stabilize the power supply in the evening during the high load period, an increase of grid integration of wind, biomass and small hydropower is necessary for the western Japan grids.

Conclusions

In this chapter, we simulated 62 GW of PV and 24 GW of wind energy integration into the Japan western grids in the high case. The Kyushu, Chugoku and Shikoku zones showed renewable electricity shares at 37%, 46% and 44% in May and at 31%, 37% and 31% in August, respectively. However, the Kansai-Chubu zone had 23% in May and 16% in August and remains far below the renewable target. As a whole, in order to achieve the 40% target, additional measures are needed.

In the high case in low load periods, around 4 to 5% of renewable electricity would be curtailed (output reduction) due to power oversupplied in Kyushu, Chugoku and Shikoku zones. A capacity development of interconnections for Kyushu to Chugoku, Chugoku to Kansai and Shikoku to Kansai would reduce excess electricity in these zones and increase renewable energy shares.

With a zero nuclear power assumption in the high case, a 15% demand decrease from the base case is required to avoid risks of supply shortage in the Kansai-Chubu zone. All zones in the western Japan grids must mobilize additional demand response measures to moderate rapid ramp-ups in residual load in the evening. Additional capacities of HPs and EVs are recommended to absorb excess electricity from PV power.

The simulation results show that inter-regional transmission and use of daytime pumping of PSHPs, HPs and EV charging accommodates around 75% of the maximum power supply from VRE in Kyushu and 80% in Chugoku in the high case.

Notes

1 METI: Long-Term Energy Supply and Demand Outlook (2015). https://www.meti.go.jp/english/press/2015/pdf/0716_01a.pdf

2 METI: Long-Term Energy Supply and Demand Outlook (2015). https://www.meti.go.jp/english/press/2015/pdf/0716_01a.pdf

3 METI: The Basic Energy Plan (2021). https://www.enecho.meti.go.jp/category/others/basic_plan/pdf/180703.pdf

4 The capacity of nuclear power plants less than 40 years would become 22.2 GW in 2030 because many of the existing nuclear reactors will exceed 40 years of age. Long-Term Energy Outlook projected the total electricity generation from all electricity sources in 2030 is 1,278 TWh. When the capacity utilization factors of nuclear generators are at

70% to 80%, nuclear electricity amount would be 136.3 TWh to 155.8 TWh. Therefore, nuclear electricity shares in the total electricity generation would be 10.7% to 12.2% in 2030.

5 METI, Agency for Natural Resources and Energy: "Recent Trends in Energy and the Progress of Policies toward Energy Transition and Decarbonization" (2019). https://www.enecho.meti.go.jp/en/committee/council/basic_policy_subcommittee/pdf/data190701.pdf

6 METI: Data from 電力調査統計 (generation business statistics), Table 1-1-2019 (generator rated capacity of generation businesses), as of March 2020.

7 METI: "Feed-in Tariff Scheme for Renewable Energy", 2012. https://www.meti.go.jp/english/policy/energy_environment/renewable/pdf/summary201209.pdf

8 Act on Special Measures Concerning Procurement of Electricity from Renewable Energy Sources by Electricity Utilities. Act No. 108 of 30 August 2011. Enforcement date: 1 July 2012. English translation is available from the following website: http://www.japaneselawtranslation.go.jp/law/detail/?id=2573&vm=&re=02

9 This highlights a clear difference from the German EEG (Renewable Energy Source Act). EEG requires TSOs and DSOs to expand the grid capacity in order to make connections of renewable energy generators to the power grids. Renewable Energy Sources Act (EEG 2017), Section 12. https://www.bmwi.de/Redaktion/EN/Downloads/renewable-energy-sources-act-2017.pdf?__blob=publicationFile&v=3

10 METI, Agency for Natural Resources and Energy: 発電設備の設置に伴う電力系統の増強及び事業者の費用負担等の在り方に関する指針 (The guideline for the expense burden of additional grid capacity development due to newly integrating generators [translated]), (2015, 2020). https://www.enecho.meti.go.jp/category/electricity_and_gas/electric/summary/regulations/pdf/h27hiyoufutangl.pdf

11 METI, Agency for Natural Resources and Energy: The guideline for the expense burden of additional grid capacity development due to newly integrating generators [translated]), (2015, 2020), Cited previously.

12 This is a significant difference from the German Renewable Energy Sources Act. EEG (2017) requires that "grid system operators must transmit and distribute renewable electricity physically, without delay and as a priority". In addition to it, these "obligations to purchase, transmit and distribute as a priority shall refer to the upstream transmission system operator". EEG 2017, Section 11 (1) and Section 11(5). https://www.bmwi.de/Redaktion/EN/Downloads/renewable-energy-sources-act-2017.pdf?__blob=publicationFile&v=3

13 Organization for Cross-Regional Coordination of Transmission Operators (OCCTO): "Outlook of Electricity Supply-Demand and Cross-regional Interconnection Lines, Actual Data for Fiscal Year 2018" (Sep. 2019).https://www.occto.or.jp/en/information_disclosure/outlook_of_electricity_supply-demand/files/190909_outlook_of_electricity.pdf

14 OCCTO: Operational Rule (2019). Article 144, "Approval of power generator or contract that require consideration of output operation". Article 144-2, "Power generators or contract subject to approval". https://www.occto.or.jp/en/about_occto/articles/files/Operational_Rules_2019.pdf

15 An example is as follow: TEPCO Power Grid: 系統アクセスルール,特別高圧編 (the grid access rule for extra high voltage) [translated], (2020). Article 6.3.11. "The transmission capacity for the generator is secured at that moment in time when a connection request is accepted" [translated]. https://www.tepco.co.jp/pg/consignment/rule-tr-dis/pdf/keitouT-j.pdf

16 METI: The Next Generation Grid Network Committee, 19th round (31 August 2020), 電力ネットワークの次世代化,系統制約の克服に向けた送電線設備の増強,利用ルールの高度化 (The next-generation power network systems, the development of transmission grids, the upgrading of grid use rules in order to overcome a limited grid

capacity) [translated]. https://www.meti.go.jp/shingikai/enecho/denryoku_gas/saisei
_kano/pdf/019_02_00.pdf

17 Enforcement Rule for the Act on Special Measures Concerning Procurement of Electricity from Renewable Energy Sources by Electricity Utilities [translated]. Article 14.8.1 to 14.8.3. (Ministerial Ordinance No.46, 2012). Note: this is an enforcement rule for the FIT law. https://elaws.e-gov.go.jp/document?lawid=424M60000400046

18 METI: 出力抑制について (Concerning curtailment on renewable electricity (feed-in management) [translated]) https://www.enecho.meti.go.jp/category/saving_and_new/saiene/grid/08_syuturyokuseigyo.html

19 In the Kyushu zone, the amount of curtailment of PV and wind electricity is 4.1% of total generation from PV and wind systems in 2019. Data from the document of Subcommittee for Basic Policies (Round 33, 17 November 2020), the Advisory Committee for Natural Resources and Energy.

20 Enforcement Rule for the Act on Special Measures Concerning Procurement of Electricity from Renewable Energy Sources by Electricity Utilities. Article 14.8.1. (amended in 2017). Cited previously.

21 In order to achieve zero-carbon emission by 2050 in accordance with the Paris Agreement of UNFCCC, the IEA Sustainable Development Scenario estimates a share of renewable-based electricity needs to be 50% of electricity generation by 2030 (https://www.iea.org/reports/world-energy-model/sustainable-development-scenario). It seems that Japan's current renewable electricity share at 19% in 2019 is too low to reach 50% by 2030. Considering these facts, our paper examines the feasibility of a renewable share at 40% in the electricity sector in 2030.

22 METI: the 18th meeting documents for "the Power Supply Network Working Group" under the Subcommittee for New Energy, 2018.

23 OCCTO: "The operational capacities for inter-regional interconnection lines for 2018–2027", 9 February 2017.

References

Kato, T., Kawai, K. and Suzuoki, Y. (2013) 'Evaluation of forecast accuracy of aggregated photovoltaic power generation by unit commitment', *IEEE Power & Energy Society General Meeting*. DOI:10.1109/PESMG.2013.6672455

Komiyama, R. and Fujii, Y. (2017) 'Assessment of post-Fukushima renewable energy policy in Japan's nation-wide power grid', *Energy Policy*, 101, pp. 594–611.

METI (2015) *Long-term energy supply and demand outlook,* (government policy document). https://www.meti.go.jp/english/press/2015/pdf/0716_01a.pdf

METI (2019) Agency for Natural Resources and Energy. 'Japan's energy 2019', https://www.enecho.meti.go.jp/en/category/brochures/pdf/japan_energy_2019.pdf

METI (2020a) 'The committee for large-scale installation of renewable energies and next generation power grid, [Translated]', https://www.meti.go.jp/shingikai/enecho/denryoku_gas/saisei_kano/pdf/020_01_00.pdf

METI (2020b) 'The Document of 'the Next Generation Grid Network Committee' (the 19th Round, 31th Aug. 2020), 'The next-generation power network systems, the development of transmission grids, the upgrading of grid use rules in order to overcome a limited grid capacity', https://www.meti.go.jp/shingikai/enecho/denryoku_gas/saisei_kano/pdf/019_02_00.pdf

OCCTO (2020) 'The Committee on Cross-Regional Network Development, 5th round, The operational capacity of interconnections (long-term plan) for 2020 to 2029' (運用

容量検討委員会,2020~2029年度の連系線の運用容量, 年間計画,長期計画), 14th February 2020, https://www.occto.or.jp/iinkai/unyouyouryou/2019/files/2019_5_1-2.pdf

Ogimoto, Y., Iwafune, K., Kataoka, T., Saito, H.A., Fukutome, K. and Isonaga, A. (2017) 'Power demand and supply analysis based on long-term energy prospects in Japan (2)', [in Japanese], *36th Meeting of Japan Society of Energy and Resources*, (conference proceedings), 28th- 29th Jan 2020, Tokyo. pp. 175–180.

Saitou, T., Urabe, C. and Ogimoto, K. (2016) 'Estimation of Japanese cumulative wind power capacity in 2050', *35th Meeting of the Japan Society of Energy and Resources*. (conference proceedings), 29th-30th Jan 2019, Tokyo.

13 A glimpse into smart cities: opportunities for the development of energy cooperatives for citizens and businesses in Mexico

Luis Román Arciniega Gil

In the last 20 years, the concept of the smart city has taken prominence both in the scientific literature and local policies, making different authors and institutions set out specific definitions and elements in order to better understand what this implies (Albino et al., 2015, pp. 3-21). Although there is no universal definition of the smart city, three documents at the international level serve as references and place the use of technologies at the heart of the concept. These include the following: the 2030 Agenda for Sustainable Development; the New Urban Agenda Habitat III – both documents from the United Nations; and the ISO 37122:2019 standards establishing definitions and methodologies for a set of indicators for smart cities. In Mexico, even though there is no normative text that homogeneously defines smart cities, the National Commission for the Efficient Use of Energy (CONUEE, 2017) has considered them to be a city model that connects its inhabitants through infrastructure and the efficient use of technology.

Smart cities are indeed a trend-setting model of local development, and while the use of technology is the main element of the concept, their wide scope allows that any local project applying innovative technologies can call itself a "smart city". However, critical literature has expressed concern about the growing role of the technological private sector in the development of smart city projects, as the marketing objectives and return on the capital of enterprises may indeed conflict with the general interest and efficient management of public resources and services, originally conceived to serve society unselfishly (Grossi & Pianezzi, 2017, pp. 79–85). In response, some authors propose developing projects with a bottom-up dimension, which implies first identifying local needs and problems, and then determining projects based on locally available resources (Niaros, 2016, pp. 51–61).

A literature review shows that clean energies are a key dimension of the smart city concept and are considered to be a viable solution for achieving integral sustainable development. Therefore, although traditionally smart cities are related to the use of technologies, this paper focuses on the manner in which locally determined energy projects can be seen as a smart solution for the development of sustainable territories. Mexico has a high potential for the development of renewable

DOI: 10.4324/9781003025962-19

energies including solar, wind, bioenergy, hydropower and geothermal energy (Elizondo et al., 2017, p. 15). Mexico's solar energy potential is particularly relevant as it has one of the top five solar irradiation levels in the world, with a daily average of 5.5 kWh/m^2 and a potential to cover up to 45% of demand (García Bello, 2019). The implementation of smart city projects, with regard to the production of renewable energies, represents an opportunity for energy transition in Mexico (IRENA, 2015) and could help to meet Sustainable Development Goal (SDG) 7 on the access to affordable, reliable, sustainable and modern energy for all, as well as SDG 11 on the creation of sustainable cities and communities.

Moreover, it is considered that the cooperative model can contribute to the sustainability of local energy projects, as it is a form of social organization motivated by the common interests of its members. In Mexico, they are defined as a

> form of organisation made up of natural persons based on common interests and the principles of solidarity, self-help and mutual assistance, with the purpose of satisfying individual and collective needs through the performance of economic activities of production, distribution and consumption of goods and services.
>
> (General Law on Cooperative Societies, Article 2)

In such considerations, cooperatives are seen, on the one hand, as entities that favor the democratic management of energy production and consumption according to the community's energy needs; and on the other hand, as entities that can contribute to bringing energy to marginal areas where investment may be costly for governments or unattractive for businesses (COPAC, n.d.).

This chapter focuses on how locally determined energy projects, combined with the cooperative model, can be seen as a smart solution from a bottom-up perspective for Mexico's national development and energy security. It starts by briefly setting Mexico's energy industry context and the country's commitments on climate change; then, it addresses the issue of energy poverty in Mexico and the challenges for the industry and business sector as leaders of energy consumption in the country; and finally, it states a conclusion that comprises recommendations and suggestions for future research, in light of Mexico's climate commitments and the new government's energy strategy (2018–2024).

Context of the Mexican energy industry

It is a matter of historical fact that in the 1930s, after a long labor dispute of "El Aguila" and others in the economic conflict with the Union of Oil Workers of the Mexican Republic (1938), the Supreme Court of Mexico ruled in favor of the workers of foreign oil companies, culminating in the expropriation and further nationalization of the oil industry (PJF, 1999). This made the energy sector largely linked to the economic performance of Mexico for the next 80 years (Colmenares, 2008, p. 53) and, for instance, between 2007 and 2013, oil revenue represented from 31 to 38% of fiscal income and from 7.4 to 8.7% of the national

GDP (Oswald, 2017, p. 155). The energy industry is a sensitive issue in Mexican politics and the participation of the private sector in this area is highly regulated by the state (Climate Transparency, 2019, p. 9).

Yet, within the last decade, Mexico has entered a period of important structural reforms that break with this long-standing tradition of a hard state monopoly on energy matters. This includes the General Climate Change Law (LGCC) of 2012; the Energy Reform of 2013; the Electricity Industry Law (LIE) of 2014; and the Energy Transition Law (LTE) of 2015. Within this framework, energy policy remains the sole responsibility of the state, but the door is open for the government to enter into contracts with individuals to participate in activities related to the electricity industry (Mexican Constitution, Article 27). This represents an opportunity to promote locally determined renewable energy projects as local governments, citizens and enterprises may participate directly in the generation, trade, transmission and distribution of electricity (CEMDA, 2017, p. 19). Moreover, the cooperative model can play a significant role in the promotion of such types of projects as it is recognized by the Mexican Constitution (Article 25) as a form of social organization for the production, distribution and consumption of socially necessary goods and services.

In that sense, and considering the solar potential that Mexico has, a strategy based on the development of photovoltaic energy could bring great benefits both for the government and particulars, thanks to the development of the cooperative model in the country. Indeed, the Ministry of Energy of Mexico (SENER, 2017) estimates that the installation of solar panels, for distributed energy generation, for 1% of the total generation capacity of the country, could represent public annual savings of around MX 1.5 billion, plus 680 million liters of unused water, and 1.3 million tons of carbon dioxide (MtCO2e) not emitted. In addition, it is considered that US$20 trillion in cooperative assets can produce around US$3 trillion of annual income (Grace et al., 2014, pp. 1–2). Mexico can then take advantage of the benefits offered by this model while diversifying its energy matrix with clean energies and increasing a culture of cooperativism in all its types, regions and social backgrounds throughout the country.

Mexico's state of play and commitments to climate change and clean energy by 2050

Mexico is one of the signatory countries of the 2015 Paris agreement (COP21), and on this basis, the country is committing itself to reduce substantially its greenhouse gas (GHG) emissions to limit the increase in global average surface temperature to below 2°C above pre-industrial levels (Bruckner et al., 2014, p. 569). In 2013, Mexico ranked 11th in the world in terms of GHG emissions, which means equivalent emissions of 489 MtCO2e, representing 1.4% of global emissions (Rodríguez, 2017, pp. 182–191). The LGCC (second transitional article) mandates to reduce the country's emissions by 30% by 2020 and 50% by 2050, in comparison with 2000; the National Congress has also set limits in the use of fossil fuels for electricity generation of 65% by 2024, 60% by 2035 and 50% by

2050. The Ministry of Energy of Mexico has integrated these objectives into the national development policies of the electricity sector, yet without precise quantitative targets due to the absence of technical and economic studies on the large-scale integration of different types of clean and renewable energies (Vidal-Amaro et al., 2015, pp. 80–96).

The latest National Inventory of GHG emissions of Mexico (2018) states that the country emitted 683 MtCO2e in 2015 of which 64% corresponded to fossil fuel consumption; 10% from livestock production systems; 8% from industrial processes; 7% related to waste management; 6% from fugitive emissions from oil, gas and mining extraction; and 5% generated by agricultural activities. Conversely, 148 MtCO2e were absorbed by vegetation, mainly forests and jungles; the balance between emissions and removals for 2015 was then 535 MtCO2e, while in 1990 these were 445 MtCO2e. This means that between 1990 and 2015, Mexico's emissions increased by 54%, with an annual growth rate (AER) of 1.7%, but also AER from 2010 to 2015 decreased to 0.8% (INECC and SEMARNAT, 2015).

In such a context, Elizondo et al. (2017, p. 19) project that by 2050, industry and transportation are likely to lead Mexican energy consumption, accounting for around 80% of demand; buildings (heating, cooling and cooking) will account for 15–20%; and the remaining 3–7% will be related to lighting. While it is estimated that half of the energy generated in 2050 will come from fossil fuels (oil and gas), it is also considered that there are good opportunities for Mexico to rely primarily on clean energy for electricity supply (mostly renewable energies) essential for heating and tackling energy poverty in the country (Elizondo et al., 2017, p. 24). Furthermore, considering that the industrial sector is the absolute leader in terms of energy demand, the government's energy strategy should consider not only the development of clean energy but also promote energy efficiency and coordinate actions with local and transport authorities in order to comprehensively address GHG reduction, whereas mitigation measures could be shared between agricultural waste management and forestry (Elizondo et al., 2017, p. 24).

The challenge of inequality: tackling energy poverty through locally determined energy projects

Although there is no harmonized definition of energy poverty in Mexico, García-Ochoa and Graizbord (2016, pp. 289–337) consider that this concept encompasses the lack of access to energy, the insufficient supply and the lack of economic capacity of households to meet all energy needs. In this regard, Mexico has 12.4 million households living in energy poverty (43.4% of the total), of which 7.8 million reside in urban areas (27.5%) and 4.5 million in rural environments (16%); additionally, energy poverty in Mexico increases as one moves from the urban to the rural, although in absolute terms, urban energy poverty is almost double that seen in rural areas (García-Ochoa, 2014, p. 19). Locally determined energy projects could represent a smart alternative to address Mexico's energy backlog and the cooperative model can effectively support the implementation of local energy projects (ILO, 2013).

Local-scale projects also represent an area of opportunity as evidence shows that large energy infrastructures may come into conflict with local populations who are affected by the implementation of major projects. This is the case, for instance, of socio-environmental conflicts related to the construction of hydro-electric plants in the states of Veracruz and Puebla (Casas Mendoza & Carbajal, 2017, pp. 70–93), as well as wind power plants in the states of Oaxaca and Yucatan (Zárate Toledo & Fraga, 2016, pp. 65–95). Considering such issues are particularly important in terms of investment risks since the cancellation of projects has represented losses of approximately MX 3 billion pesos (CEMDA, 2017, p. 33). Mexico's solar potential represents in those terms an opportunity to meet the energy needs of the poorest, as photovoltaic energy production can satisfy the demand for shorter lead times and reduce potential socio-environmental conflicts linked to other types of renewable energies (Vergara et al., 2016, p. 21).

According to the International Energy Agency (IEA), 70% of electricity access can be affordably met by renewable energy; for 65% of non-electrified households, the cheapest way could be through mini-grids and in 45% of households through off-grid technology (OECD/IEA, 2011 in ILO, 2013, p. IX). In this context, local and regional governments in Mexico could promote mini-grids and off-grid based solutions to address the country's energy backlog and basing projects on the cooperative model could help overcome the lack of institutions, policies, enterprises and human capacity that is characteristic of the most marginalized territories (ILO, 2013). Subnational governments can also take advantage of the distributed generation scheme, regulated by Electric Industry Law (Article 3 XXIII), as it promotes the capacity of local communities to propose, design, implement and operate their own technology (Dafermos et al., 2015, pp. 459–460).

In this respect, the Mexican legal framework provides that producers generating less than 0.5 MW are considered as "exempt generators", meaning that they do not require permission for energy production and may also participate in the wholesale electricity market (WEM) through the representation of a qualified service provider (LIE, Article 3 XXV). The distributed generation model in Mexico can thus enable micro-generators to meet their energy needs while facilitating the sale of surpluses in a simpler and faster manner than higher capacity sources (White & Case, 2014, p. 3). Furthermore, this type of project can activate the mechanisms considered by the Ministries of Finance and Welfare to provide clean energy to rural communities and marginalized urban areas at the lowest cost (LIE, Article 116). Projects could also benefit from the "sustainable energy financing" mechanism, considered by the Energy Transition Law (Article 55), to replace energy-inefficient equipment and devices, make improvements to buildings and install economically viable equipment to enable households to use renewable energy sources to meet their energy needs.

In summary, Mexico's energy poverty could be tackled by all three levels of government by promoting cooperative photovoltaic energy projects. In fact, electricity access can be affordably met by renewable energy and the cooperative model presents a series of competitive advantages based on a democratic control that allows better decision-making in the production, supply, distribution and consumption of electricity (ILO, 2013). Likewise, this type of project can contribute to bringing energy to marginal areas where investment can be costly for governments or unattractive for businesses (COPAC, n.d.). From this perspective, local and regional governments, in partnership with the national government, could promote community energy systems to effectively address the energy backlog in the country, particularly in marginalized urban and rural areas, while empowering people and increasing local knowledge for sustainable energy generation (Walker & Simcock, 2012, p. 194).

Reconciling interests: the benefits of the cooperative model for industry and business

Although energy production is relatively easy for "exempt generators", they do not always meet the needs of industries and businesses (leaders in energy consumption by 2050) that have to resort to medium-scale production (0.5 to 10 MW); yet the installed production capacity of small, medium and large photovoltaic arrays in recent years shows a significant growth trend that represents an opportunity for cooperativism within the Mexican industrial and business sector (DGRV, 2018, p. 122). Mexican legislation considers producers of more than 0.5 MW as "qualified generators", therefore subject to the regulations of the WEM (LIE, Article 3 XLVII). Plants between 0.5 and 2 MW are particularly affected by this legal framework as the installation process is more complex and the investment less profitable; however, it is considered that industries and businesses can benefit from the cooperative model by implementing an investment system that allows the installation of larger capacity generators under a collective measurement scheme for energy production and consumption (DGRV, 2018, p. 120).

In fact, Mexican law considers a model of isolated supply for the generation or import of electric energy to satisfy one's own needs (LIE, Article 22). The Energy Regulatory Commission of Mexico (CRE, 2017) has reinforced this criterion and provided that a group of persons that have similar commercial and financial interests can be considered a "group of economic interest" in order to meet personal energy needs. This concept is in line with Mexican jurisprudence on administrative matters, which considers that a "group of economic interest" can coordinate its activities to achieve a certain common objective (Thesis I.4o.A. J/66. *Group of economic interest. Its concept and elements that make it up in terms of economic competition.* Volume XXVIII, November 2008, p. 1244). As such, this could be seen as an opportunity for industries and businesses as it allows synchronizing their activities in order to achieve a common goal and generate more than 500 kW without carrying out all the formalities for participating in the WEM (DGRV, 2018, p. 124).

Notwithstanding, it is worth noting that financing mechanisms, especially for power plants of more than 0.5 MW and less than 2 MW, still need to be strengthened as this is the sector with the lowest return on investment (DGRV, 2018, p. 129). Commercial banking is developing financial mechanisms to meet the needs of the different groups in the distributed generation, although, it continues to face certain obstacles including price uncertainty that increases investment risk, and loans that do not give a grace period over the construction phase necessary to increase their attractiveness (DGRV, 2018, p. 129). Consequentially, the leadership of the state in the promotion, funding and integration of the concepts of ownership, design, effective participation and representation, with regard to energy production, is a necessary condition for the successful deployment of medium-scale generation plants in Mexico (Climate Transparency, 2019, pp. 6, 9).

LTE provides for financial and investment instruments and establishes that the resources for the energy transition must come from the federal expenditure budget, financial instruments available for public works and services, as well as from private contributions (LTE, Article 43). Funds are intended to attract and guide public and private financial resources, national or international, to support programs and projects that help to diversify and enrich the energy matrix (LTE, Article 48). In this context, clean energy certificates (CELs) created by the law for the Use of Renewable Energies and Energy Transition Financing (2008) can be seen as a promising mechanism, as CELs are issued by the CRE and certify the production of a certain amount of electricity as coming from clean sources (LIE, Article 3 VIII).

According to the Mexican Institute for Competitiveness (IMCO), the proper regulation of CELs can help diversify the energy mix – which had an 80% share of fossil fuels in 2015 (SENER, 2016, p. 30), while fostering a CEL's exchange market in line with the objectives of clean energy production of Mexico (IMCO, 2014, in Moreno Sánchez, 2017, p. 28). In this sense, CELs could encourage the purchase of surplus energy from medium-sized cooperative producers, thus promoting investments in energy generation, while mitigating vulnerability to supply interruptions and reducing market volatility (IPCC, 2011, p. 20). Cooperatives can then generate their own clean energy and consider a long-term repayment plan to cover the initial investment (DGRV, 2018, p. 128) through the sale of surplus energy to a trading service provider or supplier (LIE, Article 23).

While from a business perspective the objective is to make projects profitable at a low cost per kWh by competing with the prices established by the WEM, medium-scale cooperatives can offer other types of benefits to their members. This includes a reduced technology price thanks to the economies of scale, savings in electricity consumption compared with high tariffs, collective awareness raised by actively participating in the generation of clean energy and the possibility of selling the surplus generated to the distributed generation network (DGRV, 2018). In addition, they can collaterally favor access to affordable clean energy, encourage the creation of local jobs, contribute to improved performance of the national energy grid and, in general, promote economic sustainability (ILO, 2013, p. 29) and energy security for the poorest communities, industries and businesses.

Conclusions

The smart city concept has developed in recent years as a way of providing solutions to local problems through the use of technology. Although the concept is broad and its scope diverse, the implementation of clean energy projects is shown to be a central axis of the smart city concept and can contribute to promoting sustainable development and reducing carbon emissions. In the case of Mexico, history and national economic performance are strongly linked to the energy sector, which makes the participation of the private sector in this area highly regulated by the state. Furthermore, although in recent years the Mexican energy sector has been opened to various types of investments, in particular by states, municipalities and the private sector, in practice, energy policy remains a matter of federal competence, which means that there is a degree of centralization in this area that significantly affects the development of smart energy projects at the local scale.

The cooperative model can help reconcile interests within the Mexican energy sector, thus contributing to the development of local projects, the promotion of investment and diversification of the energy mix in favor of energy security and the creation of a democratic national energy system. Indeed, cooperatives are shown to be important contributors to the development of smart local energy projects, favoring democratic management of energy production and consumption according to the energy needs of a community, while bringing positive benefits to the state, business and society at large. In addition, based on the cooperative model, the development of locally determined smart energy projects can help address two main issues related to the energy sector in Mexico such as the country's energy backlog and poverty, and the reduction of carbon emissions from the industrial and business sectors as leaders of energy consumption by 2050.

In this way, the government could promote mini-grids and off-grid solutions to address the country's energy backlog, while overcoming the lack of institutions, policies, enterprises and human capacity of most marginalized areas. Micro-generators can then meet their energy needs and sell their surplus energy production under the country's distributed generation scheme. Moreover, projects can take advantage of the policies and strategies considered by the Ministries of Finance and Welfare, as well as of the "sustainable energy financing" mechanism considered by the Energy Transition Law, to provide clean energy to rural communities and relegated urban areas at the lowest cost. Finally, they can allow better decision-making in the production, supply and distribution of electricity, while empowering communities and increasing local knowledge for sustainable energy generation.

In the same manner, the government can encourage cooperativism among industries and businesses as "groups of economic interest" in order to produce their own energy needs and promote energy efficiency in this sector. The expected benefits can be translated into a reduction in the price of technology through economies of scale, savings in electricity consumption compared with high tariffs, collective awareness raised by actively participating in the generation of clean energy and the possibility of selling the energy surplus to the distributed generation network. In this context, CELs could be seen as a tool to promote a market for the purchase and sale of clean energy, enabling medium-scale cooperatives to meet their energy needs, while considering a repayment plan to cover the initial investment through the sale of energy surplus to a trading service provider or supplier. This can help increase interest in clean energy production, thereby diversifying and enriching the energy mix, mitigating vulnerability to supply disruptions, as well as reducing energy market volatility.

Finally, yet importantly, although the Mexican government (2019–2024) has stressed the importance of the development of smaller-scale and locally determined renewable energy projects, major state investments have been focused, until today, on increasing the exploitation of hydrocarbons and the renovation of the national hydroelectric system. Detailed monitoring of these plans is suggested for further research and their examination, in light of climate change commitments, is relevant to the assessment of Mexico's energy transition. Similarly, the development of cooperatives as a form of social entrepreneurship, empowerment and active participation of Mexican society in energy transition is also suggested as a topic of interest for future research.

Note: Many thanks to Natasa Kurucki, electrical engineer, as well as Christopher Fernandes, renewable energy engineer, for their valuable comments that helped prepare the final version of this article.

References

Albino, V., Berardi, U. and Dangelico, R.M. (2015) 'Smart cities: Definitions, dimensions, performance, and initiatives', *Journal of Urban Technology*, 22(1), pp. 3–21, https://doi.org/10.1080/10630732.2014.942092

Bruckner, T., Bashmakov, I.A., Mulugetta, Y., Chum, H., de la Vega Navarro, A., Edmonds, J., Faaij, A., Fungtammasan, B., Garg, A., Hertwich, E., Honnery, D., Infield, D., Kainuma, M., Khennas, S., Kim, S., Nimir, H.B., Riahi, K., Strachan, N., Wiser, R. and Zhang, X. (2014) 'Energy systems', in Edenhofer, O., Pichs-Madruga, R., Sokona, Y., Farahani, E., Kadner, S., Seyboth, K., Adler, A., Baum, I., Brunner, S., Eickemeier, P., Kriemann, B., Savolainen, J., Schlömer, S., von Stechow, C., Zwickel, T. and Minx, J.C. (eds) *Climate change 2014: Mitigation of climate change. Contribution of working group III to the fifth assessment report of the intergovernmental panel on climate change.* Cambridge University Press, Cambridge, United Kingdom, and New York, pp. 511–597.

Casas Mendoza, C.A. and Morales Carbajal, C. (2017) 'Orden simulado: hidroeléctricas, territorio y deterioro socioambiental en poblaciones totonacas y nahuas de México', *e-cadernos CES*, 28, https://doi.org/10.4000/eces.2379

CEMDA – Centro Mexicano de Derecho Ambiental (2017) 'Marco jurídico de las energías renovables en México', https://www.cemda.org.mx/wp-content/uploads/2016/06/Marco-jur%C3%ADdico-de-las-energ%C3%ADas-renovables-en-M%C3%A9xico.final_pdf

Climate Transparency (2019) *Energy transition in Mexico: The social dimension of energy and the politics of climate change* [Policy paper], https://www.climate-transparency.org/wp-content/uploads/2019/06/Energy-Transition-in-Mexico-%E2%80%93-Social-dimension-of-energy-and-the-politics-of-climate-change.pdf

Colmenares, F. (2008) 'Petróleo y crecimiento económico en México 1938–2006', *Economía UNAM*, 5(15), pp. 53–65, https://www.redalyc.org/articulo.oa?id=3635/363542896004

Comisión permanente del Honorable Congreso de la Unión (2013), Decreto por el que se reforman diversas disposiciones de la Constitución Política de los Estados Unidos Mexicanos en materia energética, Tomo DCCXXXIII No. 17, México, D.F., pp. 2–14.

CONUEE – Comisión Nacional para la Eficiencia Energética (2017) 'Fact sheet on smart cities', https://www.gob.mx/cms/uploads/attachment/file/272270/smartcity _MODIFICADA.pdf

COPAC – Committee for the Promotion and Advancement of Cooperatives (n.d.) 'Transforming our world: A cooperative 2030. Cooperative contributions to SDG7', https://www.ilo.org/wcmsp5/groups/public/---ed_emp/---emp_ent/---coop/documents/ publication/wcms_633316.pdf

Dafermos, G., Kotsampopoulos, P., Latoufis, K., Margaris, I., Rivela, B., Washima, F.P., Ariza-Montobbio, P. and López, J. (2015) 'Energía: conocimientos libres, energía distribuida y empoderamiento social para un cambio de matriz energética (v.1.0)', in Vila-Viñas, D. and Barandiaran, X.E. (eds) *Buen Conocer – FLOKSociety. Modelos sostenibles y políticas públicas para una economía social del conocimiento común y abierto en el Ecuador.* IAEN-CIESPAL, Quito, Ecuador, pp. 431–476.

Dave Grace & Associates (2014) 'Measuring the size and scope of cooperative economy: Results of the 2014 global census on cooperatives', Study prepared for the United Nations Secretariat, Department of Economic and Social Affairs, Division for Social Policy and Development, April 2014, https://www.un.org/esa/socdev/documents/2014 /coopsegm/grace.pdf

DGRV - German Confederation of Cooperatives (2018) *Potencial de las cooperativas de energías renovables en América Latina: la generación distribuida en Brasil, Chile y México.*

Elizondo, A., Pérez-Cirera, V., Strapasson, A., Fernández, J.C. and Cruz-Cano, D. (2017) 'Mexico's low carbon futures: An integrated assessment for energy planning and climate change mitigation by 2050', *Futures*, 93, pp. 14–26.

García Bello, A. (2019) 'En energía, México debe apuntar al sol', *Deloitte*, 6 August, https:// www2.deloitte.com/mx/es/pages/dnoticias/articles/energia-solar-en-mexico.html

García Ochoa, R. (2014) 'Pobreza energética en América Latina', Documentos de proyectos 576, United Nations Economic Commission for Latin American and the Caribbean (CEPAL), https://ideas.repec.org/p/ecr/col022/36661.html

García-Ochoa, R. and Graizbord, B. (2016) 'Caracterización espacial de la pobreza energética en México. Un análisis a escala subnacional', *Economía, sociedad y territorio*, 16(51), pp. 289–337, http://www.scielo.org.mx/scielo.php?script=sci_arttext &pid=S1405-84212016000200289&lng=es&tlng=es

Grossi, G. and Pianezzi, D. (2017) 'Smart cities : Utopia or neoliberal ideology?', *International Journal of Urban Policy and Planning*, 69, pp. 79–85, https://doi.org/10 .1016/j.cities.2017.07.012

Honorable Congreso de la Unión (1994)*Ley General de Sociedades Cooperativas*, DOF: 3/08/1994, Tomo CDXCI No. 3, México, D.F., pp. 19–31.

Honorable Congreso de la Unión (2008)*Ley para el aprovechamiento de energías renovables y el financiamiento de la transición energética*, DOF: 28/11/2008, Tomo DCLXII No. 19, México, D.F., pp. 88–94.

Honorable Congreso de la Unión (2012)*Ley General de Cambio Climático (LGCC)*, DOF: 6/06/2012, Tomo DCCV No. 4 México, D.F., (Segunda sección) pp. 1-29.

Honorable Congreso de la Unión (2014)*Ley de industria eléctrica (LIE)*. DOF: 11/08/2014, Tomo DCCXXXI No. 8, México, D.F., pp. 44–88.

Honorable Congreso de la Unión (2015)*Ley de Transición energética (LTE)*, DOF: 24/12/2015, Tomo DCCXLVII No. 20, México, D.F., pp. 25–53.

ILO – International Labour Organisation (2013) *Providing clean energy and energy access through cooperatives.* ILO, Geneva. ISBN 978-92-2-127528-2.

INECC – Instituto Nacional de Ecología y Cambio Climático and SEMARNAT – Secretaría (2015) *Primer Informe Bienal de Actualización ante la Convención Marco de las Naciones Unidas sobre el Cambio Climático.* INECC/Semarnat, México. https://unfccc.int/resource/docs/natc/mexbur1.pdf

IPCC – Intergovernmental Panel on Climate Change (2011) 'Fuentes de energía renovables y mitigación del cambio climático, Resúmen para responsables de políticas y resumen técnico', https://www.ipcc.ch/site/assets/uploads/2018/03/srren_report_es-1.pdf

IRENA – International Renewable Energy Agency (2015) *Renewable energy prospects: Mexico. REmap 2030 analysis.* IRENA, Abu Dhabi.

ISO – International Organization for Standardization (2019)*ISO 37122:2019 Sustainable cities and communities – Indicators for smart cities.* ISO publishing, May 2019, Geneva, 95 pages.

Mexico. Case Law [Administrative] (2008). 'Tribunales Colegiados de Circuito. Tesis: I.4o.A. J/66. Grupo de interés económico. Su concepto y los elementos que lo integran en materia de competencia económica. Novena Época, Semanario Judicial de la Federación y su Gaceta Tomo XXVIII', Noviembre de 2008, p. 1244. Registro digital: 168470.

Mexico. Court Ruling. "El Aguila" and others in the economic conflict with the Union of Oil Workers of the Mexican Republic (1938) 'La cuarta sala de la Corte niega el amparo y en parte lo sobresee a "El Aguila" y coagraviados en el conflicto de orden económico con el Sindicato de trabajadores petroleros de la República Mexicana. Sesión del 1º. de marzo de 1938', Semanario Judicial, 5a. Epoca, LV, Segunda Parte, No. 125.

Mexico. CRE – Comisión Reguladora de Energía (2017) 'Acuerdo No. A/049/2017 por el que se emite el criterio de interpretación del concepto "necesidades propias"', establecido en el artículo 22 de la Ley de la Industria Eléctrica, y por el que se describen los aspectos generales aplicables a la actividad de Abasto Aislado' DOF: 21/11/2017, Tomo DCCLXX No. 16, Ciudad de México, pp. 82-105.

Moreno Sánchez, A.L. (2017) 'The new legal and regulatory framework of the Mexican electrical sector: Possibilities of inclusion of SMEs companies', in James A. Baker III (ed.) *The rule of law and Mexico's energy reform.* https://www.bakerinstitute.org/media/files/files/c4826773/MEX-pub-RuleofLaw_ALMG-032117.pdf

Niaros, V. (2016) 'Introducing a taxonomy of the 'smart city': Towards a commons-oriented approach?', *Journal for a Global Sustainable Information Society*, 14 (1), pp. 51–61. https://doi.org/10.31269/triplec.v14i1.718

Oswald, U. (2017) 'Energy security, availability and sustainability in Mexico', *Revista Mexicana de Ciencias Políticas y Sociales UNAM*, 230, pp. 155–196.

PJF – Poder Judicial de la Federación (1999), "Decreto expropiatorio del 18 de Marzo de 1938" In *La Suprema Corte de Justicia de la Nación durante el gobierno del General Lázaro Cárdenas (1935-1940). Parte III,.* Encuadernadora progreso, México, pp. 194–195.

Rodríguez S. (2017) 'El reto del cambio climático más allá de 2018', in IMCO – Instituto Mexicano para la Competitividad (ed.) *Memorándum para el Presidente (2018–2024)*, pp. 182–191, https://imco.org.mx/indices/memorandum-para-el-presidente-2018-2024/capitulos/mexico-es-la-economia-numero-15-del-planeta-mapa-de-ruta-para-comportarnos-como-un-actor-global/el-reto-del-cambio-climatico-mas-alla-de-2018

SENER – Secretaría de Energía (2016) *Programa de Desarrollo del Sistema Eléctrico Nacional 2016–2030.* México.

SENER – Secretaría de Energía (2017) 'La Reforma Energética facilita el uso de la energía solar para pequeños generadores', 10 January 2017, https://www.gob.mx/sener/prensa/la-reforma-energetica-facilita-el-uso-de-la-energia-solar-para-pequenos-generadores

United Nations/Framework Convention on Climate Change (2015) *Adoption of the Paris agreement*, 21st Conference of the Parties. United Nations, Paris.

United Nations General Assembly (2015) *Transforming our world: The 2030 Agenda for Sustainable Development*, A/RES/10/1. United Nations, New York.

United Nations General Assembly (2016) *New urban agenda*, A/RES/71/256. United Nations, New York.

Vergara, W., Fenhann, J.V. and Schletz, M.C. (2016) *Zero carbon Latin America: A pathway for net decarbonisation of the regional economy by mid-century*. United Nations Environmental Programme, Denmark.

Vidal-Amaro, J.J., Alberg Østergaard, P. and Sheinbaum-Pardo, C. (2015) 'Optimal energy mix for transitioning from fossil fuels to renewable energy sources – The case of the Mexican electricity system', *Applied Energy*, 150, pp. 80–96, https://doi.org/10.1016/j.apenergy.2015.03.133

Walker, G. and Simcock, N. (2012) 'Community energy systems', in Smith, S.J., Elsinga, M., Fox O'Mahony, L., Seow Eng, O., Wachter, S. and Lovell, H. (eds) *International encyclopedia of housing and home*, Vol. 1. Elsevier, Oxford, pp. 194–198.

White & Case (2014) 'Reforma Energética en materia de Electricidad', *Energy, Infrastructure and Project Finance*, August 2014, https://news.whitecase.com/29/4127/downloads/09539-energy-reform-power-span-06.pdf

Zárate Toledo, E. and Fraga, J. (2016) 'La política eólica mexicana: Controversias sociales y ambientales debido a su implantación territorial. Estudios de caso en Oaxaca y Yucatán', *Trace*, 69, pp. 65–95, http://www.scielo.org.mx/pdf/trace/n69/2007-2392-trace-69-00065.pdf

Conclusion and avenues for further research

Magali Dreyfus and Aki Suwa

This volume addresses a cross-cutting exploration of the links between sustainability and local energy governance facets. The book is the cumulative research from experts from diverse disciplines to reflect on the implications of energy governance, how that can be brought into reality and what policy arrangements are required to deliver the necessary political and institutional reactions. The coverage of the compiled contributions grants us the merit to draw lessons and make recommendations for further transitions in local energy governance, reflecting on the critical arguments based on the current examples.

A similar energy policy and market structure development

Energy policy in industrial nations has long been dominated by economic motives, weighing much priority than equity and sustainability for decades. France and Japan, both industrial states, have similarities in their history of energy policy. Local authorities were initially the main actors in energy governance, supplying electricity with local resources (mostly renewable nature) to deliver basic services such as public lighting and heating to their residents. Yet before and after World War II, the strategic dimension of energy was recognized more than ever, thus local energy businesses were taken over by central governments, which organized the sector and market around a few incumbents.

The emergence of climate change as a top priority in political agendas and the pressure from international commitments accelerated the development of renewable energy. The renewable imperative has triggered a change in energy institutions and the sector and encouraged a shift in governance within which renewables are being explored. As a result, there are increasing cases in France and Japan where local actors (citizens, communities, companies) participate in the development of renewable energy production and consumption.

A slow and controlled opening of energy governance

In France as in Japan, we observe an increasing interest and involvement of local actors in the governance of energy (Suwa). This is highly related to the very nature of renewable energy, which is embedded in a territory and its physical

DOI: 10.4324/9781003025962-20

conditions. It is also notable that environmental motives seem to be in a secondary position, in comparison with socioeconomic reasons. The distributed nature
of renewable energy inevitably means energy production, which has been in the
hands of incumbent utilities with centralized production systems, is physically
dispersed and in the vicinity of the general public, who may neither be energy
nor climate experts. The not-in-my-backyards attitude (NIMBY) is the classic
example of unfamiliarity among the public toward energy production, whether
renewable or not. In that regard, socioeconomic justification is an influential logic
to overcome the cognitive and psychological barriers when facing energy development in the locality.

There is also definitely a stimulus from central governments in some countries
to promote the involvement of local actors in the energy sector. The recent legislative evolution in France (Dezobry and Dreyfus; Poupeau) clearly illustrates this.
On the other hand, the Japanese government is only partially supporting local
energy development, and most of the local stakeholders are struggling to emerge
and survive in the absence of adequate policy backing (Suwa).

Yet, a fine analysis of the reality on the ground shows that different discourses exist and that ultimately central governments remain the main actor of the
energy policy along with the traditional incumbents. This is particularly so in the
French and Japanese contexts where it is difficult to shift from traditional energy
sources because markets were tailored to their promotion (Dezobry and Dreyfus;
Takehama and Utagawa).

This means in the end that little energy transition can be expected if there
is no change in the regulatory frameworks (and, of course, sufficient political
action to enable such legal transitions). Here, despite liberalization processes and
the establishment of financial support mechanisms for renewable energy, there
appears to be much inertia, partly because of the vested interests of incumbent
market players (Poupeau). In that regard, pressure from citizens and communities over policymakers might be the key factor for a change as the Japan example
demonstrates (Iyoda).

Common multiple benefits as an incentive for local actors

Local energy governance may be promoted at a national level, but also sponta-
neous development is witnessed. These developments are mostly a response to
the various benefits designed to stimulate the relevant local implication. As the
various chapters in this book demonstrate, these benefits are diverse, overriding
economic, social and environmental factors, where the financial motive and the
revitalization of the local economy are, inter alia, a strong incentive. This seems
particularly true in rural areas, especially in Japan, where a shrinking population
needs to find new development opportunities. Benefits range from job creation,
tax income, citizen empowerment, reducing greenhouse gases emissions and so
on (Nakayama). Independence from national grids and policies, that is, local supply security, is also a goal that we see in France and Japan, as shown by the ambition to become more resilient in the TEPOS example in France (Lormeteau) or

by the 100% RE initiatives in Japan following the Fukushima disaster (Iyoda). In addition, socio-anthropological observations in Austria and Germany highlight potential social benefits, such as the emergence of a sense of identity for the community and renewed solidarity between community members who play key roles in the emergence of local energy governance (Dobigny). All these benefits contribute to the local acceptance of energy projects, which remain a key challenge in both countries as well as abroad (Kim and Lee).

Local companies, a key instrument for local actors

In order to materialize the fair and equitable distribution of such local benefits, institutional design is crucial, as it could define different forms of local partnerships, leaving a different margin of control over the activities of local actors. In France, as in Japan, there is a wide variety of legal forms, where the choice of the adequate corporate form is of utmost importance.

Against this background, the cooperative model stands out as one of the essential forms to allow citizens to be involved and take control in renewable energy projects and benefit distribution (we see this in Japan, France, Austria, Germany and Mexico). The combination of a "cooperative" company (and its quasi-forms, e.g., a special purpose corporation in Japan), rooted in equity and sharing values, and the development of renewable energy can thus constitute a good opportunity. At the same time, the training of local staff to create local expertise in the energy sector is required in the long run to sustain local energy interests (Inagaki and Nakayama; Lormeteau; Vanneaux).

This can be true in industrialized nations, but also in the most vulnerable areas in the world where local companies could provide an opportunity to raise funds, establish partnerships with energy experts from the private sector and encourage local governments to willingly engage in energy development. It is surely in line with the United Nations' Sustainable Development Goal 7 (Arciniega Gil).

Avenues for further research

More comparison

The present work, which analyzes the development of local energy governance in France and Japan, calls for more comparisons, in particular with countries considered to be well-advanced in this field. As seen in several of our chapters, countries such as Germany and Austria may provide remarkable insights. The comparative aspects may include the various legal forms of partnerships, which enable local actors to access markets and develop long-term community business activity (e.g., the *Stadtwerke* model). Another hypothesis is that the form of the states is a key factor, where federal states with a long tradition of administrative decentralization and citizen participatory processes are more likely to deliver the development of local and contextualized energy governance.

Digitalization

The digitalization of the energy sector through information and communication technologies (ICT) is already underway. Tools such as smart meters, smart grids and blockchains are already becoming part of current energy policies and infrastructure, as shown by smart cities initiatives, which now connect to the energy sector. The smart energy model would bring new big powerful actors, such as ICT companies, into the governance paradigm. This may raise a concern that local authorities might be in an unequal position, in terms of financial resources and technical expertise, with their new and dominant counterparts. In fact, technological and ICT approaches are primarily business oriented and influenced by big technological companies, whose interest is not necessary on local priorities (Languillon-Aussel). Also, the energy digitalization model remains mostly state-driven and therefore often lacks legitimacy on the ground (Arciniega Gil).

At the same time, energy digitalization also raises new challenges and opportunities for social transformation. Fast, and vast, exchange of energy supply-demand information can stimulate energy consumers' behavioral changes, as well as aggregating the bulk of demand to generate purchasing power from energy producers (Ahl et al. 2020). The connectivity to energy transaction would thus have a significant impact on both consumers and producers, while it could also open ways for citizen empowerment with possible peer-to-peer flows, facilitating the emergence of new "prosumers" (producers and consumers). Yet, they might be exclusive, as they require specific expertise and assets and will therefore need to be carefully designed.

Summary

France and Japan, as in other countries, have undergone a transformation of their energy governance systems, providing a wider scope of powers to local players, be they local governments, communities or individual citizens. This can be explained by the special nature of renewable energies, of which availability is highly related to the physical and socioeconomic contexts of local territories. In addition, increasing awareness of the multiple benefits associated with energy activities is an important incentive for local authorities, companies and citizens to act in this field. These benefits range from economic ones to social and cultural ones. Advantages of a more political nature are also significant and need to be considered when planning energy projects, no matter what motivations may be, for example, democratic aspirations of some communities or ambitions to be more resilient and autonomous in terms of security of supply in the face of disasters.

The conclusions drawn from the different chapters, however, show that local governance of energy is still in an emerging stage. In these two countries, which are traditionally very centralized, path dependency and a clear intent from central governments to keep decision-making power have a strong influence upon local actors, who are seen as potential partners or executants for the state rather than real peers with which a territorial energy policy could be designed. This is not surprising given the real strategic dimension of energy policy at the national and international scales, yet it shows a kind of doublespeak attitude from the governments. On the one hand, the need to involve local actors is acknowledged, while on the other hand, they are not given sufficient means to develop their activities.

Yet institutional inertia means that the process might be too long with regard to the urgency of reducing greenhouse gas emissions and adapting to climate change impacts. This is particularly clear in the very slow changes made to regulatory frameworks. Despite new legal opportunities in forming renewable energy communities, or new public-private companies for energy purposes, older rules prevent the penetration of renewable energy and new entrants into the market and grids, to the benefit of the main incumbents (traditionally in a position of monopoly and producing fossil or nuclear energy) who remain key spokespersons for the governments.

Against this background, the landscape of the energy sector emerging from this book shows the plurality of actors and levels of intervention that need to be coordinated if renewable energy is really to develop. In that regard, the increased awareness and empowerment of citizens and communities is a sure way to accelerate the implementation of policies pursuing that goal.

This book addresses the need for effective governance, accounting for different dimensions and actors having strong implications on the sustainable energy transition. In many countries, the problem is the recognition and engagement of stakeholders, which ultimately determines to what extent vital actions are to be taken to materialize the sustainable energy transition.

This consecutively sheds light on the next research agenda for local energy governance, which demands further inquiry as to how to ensure the emergence of stakeholder engagements and how to keep the momentum for a sustained period. Delivering such scrutiny would promise the beginning of an agenda for sustainability in the local energy governance domain and a solid foundation for a fair and equitable future.

Index

Note: Page numbers in *italics* indicate figures, **bold** indicate tables in the text, and references following "n" refer endnotes

Abe, S. 37
"Accelerating the energy transition" 114
ACER Regulation (EU) 2019/942 29n2
ACIC *see* agricultural collective interest companies (ACIC)
"Act on Special Measures Concerning Procurement of Electricity from Renewable Energy Sources by Electricity Utilities" 219–220
ADEME 52, 53, 157n11, 206
ADLC *see Autorité de la Concurrence* (ADLC)
AEMS *see* Area Energy Management System (AEMS)
"affordable and clean energy" 161
Agency of Natural Resources and Energy (ANRE) 33
Agenda for Sustainable Development, 2030 245
Agenda 21 of the European Union 200
agreeing parties: before implementing EIA 91–93; stage of implementing EIA 95; stage of implementing SEA 94
agricultural collective interest companies (ACIC) 104
"alternative decentralizers" network of actor 56
ANAH 52
ANRE *see* Agency of Natural Resources and Energy (ANRE)
Area Energy Management System (AEMS) 206
ARENH *see* Regulated Access to Nuclear Electricity (ARENH)
Association of French Mayors 57
Austria 181–182; circumstances and actors of 183–184; economic benefits and energy tourism 188–190; local RE transition, motives of 184–188; methods and case studies 182; new solidarities 190–191; recognition, identification and collective identity 191–193
Autorité de la Concurrence (ADLC) 23, 24

barriers to renewable energy 85; case record 90–96; environmental impact assessment 88–90; local autonomy system in South Korea 86; RPS *vs.* FIT systems 87–88, **88**
baseload electricity 23, 24
baseload hydro-electrical plants 23
Basic Act on Energy Policy 33
Basic Energy Plan in 2010 214–216
Batty, M. 199
BDH *see* biomass district heating (BDH)
benefit sharing 63, 64; and local energy governance 66–68
Biden, J. 37
'Big Four' energy companies 181, 184
biomass 26, 49–50; biomass district heating (BDH) 182, 185, 187, 192; energy demands 39; heat supply business 169–172; woody biomass 163, 173, 174, 176–178
Bomi Kim 7
Bouillon, C. 154
Bush, G. W. 35
business model of heating networks 49

capital-intensive purposes 5
CASA *see* Citizens' Alliance for Saving the Atmosphere and the Earth (CASA)

CCREPP *see* Citizens' Co-owned Renewable Energy Power Plants (CCREPP)
CEC *see* climate and energy contribution (CEC)
CELs *see* clean energy certificates (CELs)
CEMS *see* community energy management system (CEMS)
centralization of urban information systems 202
centralized energy system 4
centralized governance 20–25
centralized market structure 64, 79
centralized top-down governance scheme 25
CEPFEP *see* Chuo Electric Power Furusato Heat and Power (CEPFEP)
Champsaur Commission's report 24
Chiba University of Commerce (CUC) 38
Chugoku zone, island of Japan: CO_2 emission levels in 236; grid balance in *232*, 238; power oversupply in 233, 241; power transmission from *234*; renewable electricity shares 239; simulation results in 237, 240; VRE power in 235, 241
Chuo Electric Power Furusato Heat and Power (CEPFEP) 75–76
"Citizen Energy Communities" 19–20
Citizens' Alliance for Saving the Atmosphere and the Earth (CASA) 35
Citizens' Co-owned Renewable Energy Power Plants (CCREPP) 39
civil society: actions, to promote renewable energy 37–40; advocacy and movements 41; and associated organizations 32; climate and energy issues, advocacy on 35–37; and progressive actors 40
clean energy certificates (CELs) 251–252
"Clean Energy for all Europeans Package" 17, 19, 29n2, 49
clean energy production 253
CLER (Network for Energy Transition) 146
Climate Action Network Japan (CAN-Japan) 38
climate and energy contribution (CEC) 57
Climate and Resilience Law of 2019 26
climate institutional framework: European local energy governance, new actors in 19–20; liberalization of energy markets 17–18, *18*; renewable energy and energy transition, promotion of 18, **19**

Climate Week NYC 38
Clinton, B. 199
CMS *see* community management system (CMS)
coal 4, 7, 25, 26, 33, 36, 37, 40, 49, 70, 185, 219, 221, 223, 225, 228
collective services schemes (SSC) 54
community-based organizations 32
community-based renewable electricity developments 77
community business entities 75–76
community economic benefits 81, 124
community energy: development of 3, 63; schemes in Japan *76*; sector 122–124
community energy management system (CEMS) 207
community management system (CMS) 208
complex compensation system 204
comprehensive governance schemes 2
concession system 48, 49, 70, 104, 105, 113, 114
Consotab system 208
Constitution of the Republic of Korea 86
conventional means of production 23 cooperatives 8; community business entities 75; companies 113, 259; energy cooperatives 8, 63, 70, 78, 123, 186; entities 78; governance of projects 151; Japanese cooperatives 73; local farmers in 183–184; model for industry and business 73–75, 80, 188, 246–252, 259
cooperative societies of collective interest (CSCI) 113
Coordinated by the Climate Group 38
COP3 *see* third Conference of Parties (COP3)
CSCI *see* cooperative societies of collective interest (CSCI)
CUC *see* Chiba University of Commerce (CUC)
curtailment (output reduction) of renewable electricity 223–224

Democratic Party of Japan (DPJ) 36–37
Denjiren *see* Federation of Electric Power Companies of Japan (FEPC)
de Rugy, F. 154
Dezobry, G. 6
digital/energy transition in French cities 199–200; Japanese smart communities and 204–209; multi-scalar governance and life-cycle approach of 209–210; turn

of "urban" energy in France 203–204; urban strategies 200–204

digitalization of energy sector 260

distributed renewable energy 161

Dobigny, L. 9

DPJ *see* Democratic Party of Japan (DPJ)

Dreyfus, M. 6

eco-energy tourism 188–190

ecological dimension of RE transition 186–187

ecological transition contract 155

economic regulatory framework 20, 23–24

EDF *see Electricité de France* (EDF)

EDF-GDF *see Electricité de France-Gaz de France* (EDF-GDF)

EIA system *see* environmental impact assessment (EIA) system

"El Aguila" labor dispute of 246

Electric Industry Law 249

Electricité de France (EDF) 4, 18, 22, 24, 53, 54, 56, 59, 103, 105

Electricité de France-Gaz de France (EDF-GDF) 104

Electricity Directive (EU) 2019/944 29n2

Electricity Industry Law (LIE) 247

Electricity Regulation (EU) 2019/943 29n2

Elizondo, A. 248

emergence of climate change recognition 257

empowerment process of RE transition 184–185

Enedis 103, 111

Energy and Climate Law 151

energy citizenship 149

Energy Code 103

"energy commons" 68

"energy" cooperative model 74

Energy Efficiency Directive (EU) 2018/2002 29n2

energy governance 209, 260; effective decentralization of 152; European local energy governance 19–20; genuine decentralization of 153, 156; implications of 257; local energy governance 2–4, 6, 9, 40, 145, 147 *see also* local energy governance in Japan; methodology in 152; new local actors in 26–28; slow and controlled opening of 257–258

Energy Industry Act (1988) 140n2

energy market regulation in France 47–48; emerging players but with limited powers 50, *51*; from local to national management 48–49; stronger local influence, preservation of 49–50

energy markets: in Japan 68; liberalization of 17–18, *18*, 48

Energy Performance in Buildings Directive (EU) 2018/844 29n2

energy policy 33; and market structure development 257

energy poverty in Mexico 248–250

Energy Reform of 2013 247

Energy Regulatory Commission of Mexico 250

energy-related benefits 79

energy transition 2; civic service projects on 153; decentralization of 157; digital and *see* digital/energy transition in French cities; EU energy transition 18, **19**; issue of 152; in Japan *see* energy transition in Japan; in Mexico 246, 253; municipal role for 123; renewable energy and 18; resources for 251; sustainable energy transition 261

energy transition contracts (ETC) 155

energy transition in Japan 32; civil society's actions, to promote renewable energy 37–40; civil society's advocacy, on climate and energy issues 35–37; governmental preferences and political background 33–34; historical background and policy development 68–6709; national government, changes within 34–35; renewable policies 70–72

Energy Transition Law (LTE) 247, 249, 252

Energy Transition Law for Green Growth of 17 August 2015 106

Environmental Impact Assessment Act 89

Environmental Impact Assessment Council 95

environmental impact assessment (EIA) system 88–90; before implementing 91–93; consultation authority 97; stage of implementing 95–96

"environmentally conscious" product supply chains 74

Environmental Model Cities and Biomass Town projects 173

"Environmental Model City" 163

essential facility owner's competitors, financial strength of 24–25

ETC *see* energy transition contracts (ETC)
EURATOM 28n1
European Clean Energy Package 27
European Commission 17, 21, 114
European energy: European local energy governance, new actors in 19–20; law 19; liberalization of energy markets 17–18, *18*; renewable energy and energy transition, promotion of 18, **19**; single market 17–18
European institutional framework 28n1
European local energy governance, new actors in 19–20
European model of sustainable city 200
European Union (EU) 17, 27, 117; ADEME and 52; Agency for the Cooperation of Energy Regulators (ACER) 29n2; Agenda 21 200; energy transition targets **19**; European Commission and 17; European public policies as 18; and French law 119n49; key role of 6; liberalization process by 27; regulatory framework 17; Treaty on the Functioning of the European Union (TFEU) 114
Evrard, A. 181
"experience-oriented tourism" 190
"expert-oriented energy tourists" 189
"expert-oriented tourism" 190

feasibility of 100% renewable energy, in Japan: analytical model of local value added 162–163; business, local economy through 161–162; of local authority 172–176; by local value-added analysis model 163–172; *see also* Japan
Federation of Electric Power Companies of Japan (FEPC) 33
feedback loops 202
feed-in premium (FiP) system 71–72
feed-in tariff (FIT) system 71–72, 85, 161, 167, 213–214, 219; for renewable energy diffusion 36; *vs.* renewable portfolio system (RPS) 87–88, **88**
FEPC *see* Federation of Electric Power Companies of Japan (FEPC)
"FIT law" 219–220, 223–224
fossil fuel energy sources 1, 2, 33, 157n1
Framework Act on Environmental Policy (2006) 93
France/French: authorities 6, 17, 25, 28, 103; baseload market 24; cities' energy planning 208–209; economic analysis of nuclear power in 25; electric generation mix in 2019 25–28, *26*; *Electricité de France* (EDF) 4; electricity ecosystem 106; energy law 20, 119n49, 147, 149; energy sector 17, 47, 59, 106; Energy Transition for Green Growth law 5; legal context 26–28; legal framework 7; local authorities and energy in *see* local authorities and energy in France; local initiatives 146; local public companies and electricity *see* French local public companies and electricity; local public enterprises 8; market 21, 23, 24; model 4; nuclear industry 21; nuclear power plants 20, 21, 24; Nuclear Safety Agency 30n24; policymakers 28; production mix 49; renewable energy sources (RES) 5; searching for alternatives *see* searching for alternatives in France; socio-technical system 25; strategy 50; "TEPOS" 8; territorial organization in **10**; urban model 199; voluntary initiatives and decentralization *see* voluntary initiatives and decentralization, French energy system; *see also* energy market regulation in France
Freiamt (Baden-Württenberg, Germany) 183–185, 189
French local public companies and electricity 103–106; with majority public shareholdings 106–110; with minority public shareholdings 110–114
Fridays For Future Kyoto, climate march for 41
FSL 53
Fukushima Daiichi Nuclear Power Station accident 5, 32, 35, 36, 69, 122, 205, 209, 214–217, *216*, **217**
functional decentralization 157

García- Ochoa, R. 248
Garorim Bay Management Area Plan 93
Garorim Bay tidal plant project in South Korea 85; case record 90–96; environmental impact assessment 88–90; RPS *vs.* FIT systems 87–88, **88**
Garorim Tidal Power Plant Consortium 91, **92**
gas network 26, 42n5; electricity and 47–50, 53, 54; greenhouse gas (GHG) emissions *see* greenhouse gas (GHG) emissions

GDF 54
General Climate Change Law (LGCC) 247
German models 123
German Renewable Energy Sources Act 242n12
German taxation system 162
Germany, renewable energy territories in 181–182; circumstances and actors of 183–184; economic benefits and energy tourism 188–190; local RE transition, motives of 184–188; methods and case studies 182; new solidarities 190–191; recognition, identification and collective identity 191–193
GHG emissions *see* greenhouse gas (GHG) emissions
Global Action Plan in 2019, SDGs 1
Governance of the Energy Union Regulation (EU) 2018/1999 29n2
"government through experimentation" accords 203
GPN *see* Green Purchasing Network (GPN)
Graizbord, B. 248
Great East Japan Earthquake 5, 36, 70
"green growth" 153
greenhouse gas (GHG) emissions 1, 2, 5, 6, 17, 18, 25, 33, 55, 87, 205, 247
Green Purchasing Network (GPN) 38
"Grenelle laws" 53, 103, 105, 150
grid development plan of interconnections **224**, 224–225
grid expansion and development, for renewable businesses 219–221, **222**
grid parity achievements *72*
grid rules (grid operational rules) for interconnection use 223
grid voltage levels and renewable energy integrations, in TSO zones **222**
"groups of economic interest" 250, 252
Guettier, C. 152
Güssing (Burgenland, Austria) 183, 188, 189–192

Hasegawa, K. 27
heat, local value added by 175–176
heating mode, HPs in 231, 233
heating networks 48–50, 105
heat supply: business, biomass 169–173; from renewable energy 174–175, 178
Heinbach, K. 133
HEMS *see* home energy management system (HEMS)

High Level Political Forum 1
Hioki Energy Co., Ltd, case study 135–137, *136, 138*
historic centralized model 56
"historic Jacobins" network of actor 56, 57, 59
home energy management system (HEMS) 134
hydropower semi-public companies (HSPC) 110, 113–114

ICT *see* information and communication technologies (ICT)
IEA *see* International Energy Agency (IEA)
IEC *see* International Electronic Commission (IEC)
"Iida City Ordinance for Building Sustainable City through Introduction of Renewable Energy" 39
IMCO *see* Mexican Institute for Competitiveness (IMCO)
"Implicit Auction Scheme" for interconnection use 223
Inagaki, K. 8
independent power producers (IPP) 4
industry and business, cooperative model for 250–251
information and communication technologies (ICT) 260
initial typology, of local energy arrangement *81*
"Innovative Energy and Environment Strategy" 36–37
innovative rules of governance 147–149
interconnections, rules for 223
inter-governmental conflict 97–98
International Electronic Commission (IEC) 207
International Energy Agency (IEA) 1, 249
inter-regional transmission, utilization of 233–236, *236, 235*
IÖW model 162
IPCC 5th assessment report 32
IPCC Special Report on 1.5°C of global warming 32
IPP *see* independent power producers (IPP)
ISC *see* Izumisano City Government (ISC)
Iyoda, M. 7
Izumisano City Government (ISC) 77
Izumisano Electric Company (Izumisano PPS) 77–78, **78**, *79*
Izumisano Power Producing and Supplying Association 77, *79*

Japan: customers' conservatism 70;
electricity market reform *70*; electricity
source transition *69*; energy transition in
see energy transition in Japan; feasibility
of 100% renewable energy *see*
feasibility of 100% renewable energy, in
Japan; historical context 4–5; industry
groups 7; local energy governance in
see local energy governance in Japan;
local renewable energy development in
7; MPSs' development 122, **124–130**;
municipalities 123; renewables in
electricity production 5–6; rural
areas 66; smart community 205–206;
territorial organization in **11**; *see also*
renewable energy development in Japan;
value added to local economies in
Japan; western Japan
Japan Business Federation
("Keidanren") 33
Japan Power Exchange (JPEX) 77
Jenniches, S. 132
Jiminto 36
Johnson, C. 210n2
Jühnde (Lower Saxony, Germany) 183,
186–189, **187**, 191, 192

Kan, N. 36
Kannen, A. 192
Kansai-Chubu, island of Japan: CO_2
emission levels in 236, 241; grid balance
in *233*, 239, *239*; power transmission
from *234*; renewable electricity shares in
239, 241; simulation results in **237**, **240**;
supply capacity in 239
Kansai Electric Co., Ltd 77
keiretsu 204
KEPCO *see* Kyushu Electric Power
Company (KEPCO)
Kiko Forum 35, 43n7
Kiko Network 35–36
Konan Ultra Power scheme *74*
Kono, T. 34, 42n6
Korea Environment Institute 94
"Kyoto City Ordinance of Global
Warming Counter Measures" 40
Kyoto Protocol 35, 36, 41
Kyushu, island of Japan: case in 75–76;
CO_2 emission levels in 236; grid
balance in *232*, 238, *238*; LFC control
reserve activations in 236, *236*; power
oversupply in 233, *235*, 241; power
transmission from *234*; renewable

electricity shares 239; simulation results
in **237**, *240*; targets of high case in 228,
230; VRE power in 235, *235*, 241
Kyushu Electric Power Company
(KEPCO) 42n5, 75

Languillon-Aussel, R. 9
large-scale conventional power plants
221, 223
"late-come" renewable generators in
interconnection use 223
Law ETGG 150–152
Law on Energy Transition for Green
Growth (Law ETGG) 26, 146
LDCs *see* local distribution companies
(LDCs)
LDP *see* Liberal Democratic Party (LDP)
Lee Myung-bak 85, 87
LFC control reserve: activations 236,
236; capacity constraints 227; scheme
230–231
LGC *see* Local Governments Code (LGC)
LGCC *see* General Climate Change Law
(LGCC)
Liberal Democratic Party (LDP) 36
liberalization of energy markets 5, 17–18,
18, 48
LIE *see* Electricity Industry Law (LIE)
life-cycle approach of French smart cities
209–210
Lisbon Treaty 28n1
local actors: incentive for 258–259; key
instrument for 259
local authorities and energy in France 47;
decentralization, under strong political
constraints 56; local governance, as
source of rivalry and fragmentation
57–58; regulation of *see* energy market
regulation in France; and sectoral
policies 50–55; seeking financial and
human levers 57
local autonomy system in South Korea 86
local cyclical economic model 177
local distribution companies (LDCs)
103, 105
local energy arrangement, initial typology
of *81*
local energy governance 258; as source of
rivalry and fragmentation 57–58
local energy governance in Japan 63–64,
259; benefit sharing and local energy
governance 66–68; community business
entities 75–76; cooperative model

73–75; definition of 2–3; Japanese energy transition, strategic challenge of 68–72; municipal energy utilities 77–78; national wealth and resource security 64–66

Local Governments Code (LGC) 105

local income deficit *67*

local public electricity companies 104; with majority public shareholdings 106–110

local public enterprises (LPEs) 103–104, 106–107, 110, 112, 115–116

local RE transition 181, 193; ecological dimension of 186–187; empowerment process 184–185; local stakeholders and economy 185–186; social, fair and democratic values 187–188, **187**

local-scale projects 249

local stakeholders and economy 185–186

local value-added analysis model 163, 176

long-term electricity supply and demand plan 91

"long-term fixed generators" 223

Long-Term Supply-Demand Energy Outlook 213–215

Lormeteau, B. 8

LPEs *see* local public enterprises (LPEs)

LTE *see* Energy Transition Law (LTE)

Lyon smart community program 206–208

"MAKE the RULE" campaign 36

"management of unpaid bills" 53

MAPTAM law of 2015 54, 59

Matlab Optimization Tool Box 226

Mexican Institute for Competitiveness (IMCO) 251

Mexico 245–246; context of energy industry 246–247; cooperative model for industry and business 250–251; energy poverty in 248–250; energy sector 252; energy transition 253; legal framework 249; national development and energy security 246; state of play 247–248

microgenerators 252

Ministries of Finance and Welfare (Mexico) 249, 252

Ministry of Defense (MOD), Japan 34

Ministry of Economy and Finance (France) 57

Ministry of Economy, Trade and Industry (METI), Japan 33, 34, 205, 214, *216*, 221

Ministry of Energy (Mexico) 247, 248

Ministry of the Environment (MOE), Japan 33, 34, 37, 42n6, 43n6, 89, 90, 94, 96

Ministry of the Foreign Affairs (MOFA), Japan 34, 42n6, 43n6

Minshuto 36

Miyama Smart Energy, case study *134*, 134–135, *135*

mobility issue, in France 55

"model" town of RE transition 191

mode of citizen participation 149

"moderate decentralizers" network of actor 56

Morotomi, T. 123

MPSs *see* municipal power suppliers (MPSs)

multiannual energy programming (PPE) 55

multi-scalar governance of French smart cities 209–210

municipal business models 123

municipal energy utility model 77–78, 80

municipal power suppliers (MPSs) 122; case studies of 134–137; typology of 122, **124–130**; value added to local economies 132–133

Mureck (Styria, Austria) 183, 185–187, 189, 191, 194

Nakayama, T. 8

National Commission for the Efficient Use of Energy 245

National Energy and Climate Plan (NECP) 29n2

National Environmental Policy Act (NEPA) 88

National Forum for CCREPP 39

National Inventory of GHG emissions of Mexico 248

Nationalization Law 104

National Low Carbon Strategy 157

natural gas-powered vehicles (NGVs) 55

NECP *see* National Energy and Climate Plan (NECP)

NEDO *see* New Energy and Industrial Technology Development Organization (NEDO)

NEPA *see* National Environmental Policy Act (NEPA)

neutral parties: before implementing EIA 93; stage of implementing EIA 96; stage of implementing SEA 94–95

new and renewable energy (NRE) 87;
weighted values of **89**
New Energy and Industrial Technology
Development Organization (NEDO)
206, 208
"New Energy Vision for the
Community" 171
new organizational archetypes in Japan
72–73; community business entities 75–
76; cooperative model 73–75; forming
of municipal energy utilities 77–78
New Urban Agenda Habitat III 245
Next Generation Energy and Social
System 205
NGOs *see* non-governmental organizations
(NGOs)
NGVs *see* natural gas-powered vehicles
(NGVs)
Nikkei Telecon 21 database *41*
NND *see* non-nationalized distributors
(NND)
"NOME Act" 22
NOME Law 106
non-governmental organizations (NGOs)
32, 35–38, 90, 94, 96
non-nationalized distributors (NND) 104
non-payment and fight against fuel poverty
52–53
non-steerable renewable energies 23
not-in-my-backyards attitude
(NIMBY) 258
notorious Japanese pollution incidents 68
NOTRe Act 54
NOTRe law 52, 59
NRE *see* new and renewable energy
(NRE)
nuclear and renewable energies, in METI
214, *216*, **217**
nuclear hegemony 20–25

OECD *see* Organisation for Economic
Co-operation and Development (OECD)
Okinawa Electric Power Company 33
omega of French smart city 199–200
OPAH 52
OPATB 52
"Open ADR" 207
"operation/maintenance and business
management phase" 163–164, *167*
opposing parties: before
implementing EIA 93; stage of
implementing EIA 95–96; stage of
implementing SEA 94

Organisation for Economic Co-operation
and Development (OECD) 65
organizational and ownership governance
archetypes 64
Organization for Cross-Regional
Coordination of Transmission Operators
(OCCTO) grid development plan for
2029 **224**, 224–225

Paris Agreement on Climate (2015) 1, 5,
17, 32, 64, 247
Paris Treaty 28n1
PCAETs *see* territorial climate, air and
energy plans (PCAETs)
PCETs *see* territorial energy-climate plans
(PCETs)
PEIC *see* public establishments for
intermunicipal cooperation (PEIC)
pioneering TEPOS territories 151
"Place Au Soleil" plan 151
planned housing improvement
operations 52
"planning/introduction phase" 163, 164
Pluriannual Energy Program 26
Population Decline Committee of the
Japan Policy Council 173
Porter, M.E. 162
"positive energy territory" 150, 152, 154
Poupeau, F.-M. 7
power generation in Japan, trends of *71*
power grid systems, renewable energies
and grid rules for 217, **218–219**
Power Producer and Supplier (PPS)
40, 70
power retail market in Japan 139n1
"Power Shift" campaign 40
power sources, for Japanese MPSs 131
PPE *see* multiannual energy
programming (PPE)
PPS *see* Power Producer and Supplier
(PPS)
pressurized water reactors (PWRs) 21
private renewable energy companies
111–113
"progress asymptotes" 204
PSHPs *see* pumped-storage hydropower
systems (PSHPs)
public electricity distribution service
58, 104
public establishments for intermunicipal
cooperation (PEIC) 115
Public Finance Law (PF Law) 154
public minority shareholdings 111–113

public procurement law 119n49
public shareholding in SPCs 107, *108*
pumped-storage hydropower systems (PSHPs) 231
PWRs *see* pressurized water reactors (PWRs)

qualitative methods 182

ramp up/down rate constraints 228
RE Action 38, 41
RECs *see* Renewable Energy Communities (RECs)
regional climate air energy schemes (SRCAE) 54, 55, 157n11
regional schemes for the planning and sustainable development of territories (SRADDET) 54, 55, 57, 59, 157n11
Regulated Access to Nuclear Electricity (ARENH) 22, 23
regulatory framework, of renewable energies and grid rules 217, **218–219**
"RE100" renewables pledge in Japan 38, 41, 42n6
renewable electricity shares: in high load period *239*, 239–241, **240**; in low load period 236–238, **237**
renewable energies: capacity in Japan, trends of *73*; community 49; deployment 71; electricity from 173–174; and energy transition, promotion of 18, **19**; and grid rules, for power grid systems 217, **218–219**; heat supply from 174–175; local authority 175–176; production project 107; self-sufficiency 182
renewable energy business: biomass heat supply 169–172; electricity generation 164–169; grid expansion and development for 219–221, **222**; local economy through 161–162; at village level 163–164
Renewable Energy Communities (RECs) 19–20, 28, 87; weighted values of **89**
renewable energy development in Japan: curtailment (output reduction) of renewable electricity 223–224; electricity generation 212–214, *213–215*; Fukushima nuclear accident, nuclear and renewable energies and 214–217, *216*, **217**; grid development plan of interconnections **224**, 224–225; grid expansion and grid

development for renewable businesses 219–221, **222**; transmission grids and interconnections 223; transmission grid systems from generation businesses 217, **218–219**, 219; *see also* feasibility of 100% renewable energy, in Japan; Japan
Renewable Energy Directive (EU) 2018/2001 29n2
Renewable Energy Platform Japan 38
renewable energy territories, in Austria and Germany 181–182; circumstances and actors of 100% RE territories 183–184; economic benefits and energy tourism 188–190; local RE transition, motives of 184–188; methods and case studies 182; new solidarities 190–191; recognition, identification and collective identity 191–193
Renewable Portfolio Agreement 87
Renewable Portfolio Standards (RPS) 71
renewable portfolio system (RPS) 85, 98; *vs.* feed-in tariff (FIT) system 87–88, **88**
rent-a-roof business 168
Réseau de Transport de l'Electricité (RTE) 103
RICOH 38
Rio Earth Summit 200
Risk Preparedness Regulation (EU) 2019/941 29n2
RPS *see* Renewable Portfolio Standards (RPS)
RTE *see* Réseau de Transport de l'Electricité (RTE)
Russian Federation 36

scheduled building thermal improvement operations 52
"scheduled power supply" 223
Schreurs, M.A. 37
searching for alternatives in France 17; centralized governance and nuclear hegemony 20–25; energy sources and decentralized governance 25; European energy and climate institutional framework 17–20, *18*, **19**; French energy mix 25–28, *26*
sectoral policies 50; energy and mobility issues 55; non-payment and fight against fuel poverty, aid for 52–53; thermal renovation of buildings 50, 52; from urban planning to energy-climate schemes 53–55

self-consumption operation 27
semi-public companies (SPCs) 104,
    106–108, 115, *116*
Shiga model 74
Shiga Prefecture in 1998 39
Shikoku, island of Japan: CO_2 emission
    levels in 236, 241; grid balance in
    232; power transmission from *234*;
    simulation results in **237**, **240**; VRE
    power in 235
Shimin Forum 35
Shinada, T. 27
single centralized system 202
Single European Act in 1997 29n1
single-purpose semi-public companies
    (SPSPCs) 104, 106, 110–111, 114, 115
singular public participation company
    113–114
small-scale hydroelectric power generation
    164–169, *165*, *166*
smart city model 9
"smart communities" program 205
smart energy model 260
"smart revolution" Japan 209
sobriété 103
social and democratic values, of RE
    project 187–188, **187**
social implication of energy appropriation
    181–182; circumstances and actors of
    100% RE territories 183–184; economic
    benefits and energy tourism 188–190;
    local RE transition, motives of 184–188;
    methods and case studies 182; new
    solidarities 190–191; recognition,
    identification and collective identity
    191–193
socio-technical systems 4, 203
solar power generation business
    164–169, *169*
South Korea, Garorim Bay tidal plant
    project in 85; case record 90–96;
    environmental impact assessment
    system 88–90; local autonomy system
    in South Korea 86; RPS *vs.* FIT systems
    87–88, **88**
SPCs *see* semi-public companies (SPCs)
SPSPCs *see* single-purpose semi-public
    companies (SPSPCs)
SRADDET *see* regional schemes for the
    planning and sustainable development
    of territories (SRADDET)
SRCAE *see* regional climate air energy
    schemes (SRCAE)

SSC *see* collective services
    schemes (SSC)
Stadtwerke model 123, 138, 139, 259
stakeholders 1–6, 18, 42, 54, 64, 72;
    attitudes and relationships, of EIA 91–
    96; of business entity 80; in ecological
    transition 155; and economy 185–186;
    inhabitants and 192; institutions and
    policy actors 85; Izumisano investment
    from *78*; in Japan 78; municipal
    stakeholders 123; origins and 122;
    recognition and engagement of 261;
    spectrum of 80; transformations in 202;
    value chains of 177; winning strategy
    among 90
state-centered decisions 68
strategic dimension of energy 257
strategic environmental impact assessment
    (SEA) 90, 93–97
Suga, Y. 34
supply–demand balances of power grid,
    in 2030 212; constraints 226–228;
    control reserve activations in low
    load period 236, *236*; curtailment
    (output reduction) of renewable
    electricity 223–224; development plan
    of interconnections **224**, 224–225;
    expansion and development for
    renewable businesses 219–221,
    222; formulation 226; Fukushima
    nuclear accident, nuclear and
    renewable energies and 214–217,
    216, **217**; in high load period *238*,
    238–239; historical trends in electricity
    generation 212–214, *213–215*; in low
    load period 231–233, *231–233*; points
    for analysis 225; renewable electricity
    shares in high load period *239*, 239–
    241, **240**; renewable electricity shares
    in low load period 236–238, **237**; target
    capacity and assumptions for 228, **229**,
    230, 230–231; transmission grids and
    interconnections 223; transmission grid
    systems from generation businesses
    217, **218–219**, 219; UC-ELD model
    225–226; utilization of inter-regional
    transmission 233–236, *236*, *235*
Süsser, D. 192
sustainable electricity production 203
"sustainable energy financing"
    mechanism 252
"system operator phase" 163
"system production phase" 163

Takehama, A. 9
target capacity and assumptions, for high renewable case 228, **229**, **230**, 230–231
TECV law 54
TEPCV 146, 152–154, 156
TEPOS 146–152, 156, 258
territorial climate, air and energy plans (PCAETs) 54–55, 57, 59, 148
territorial energy-climate plans (PCETs) 54
territorial organization: in France **10**; in Japan **11**
TFEU *see* Treaty on the Functioning of the European Union (TFEU)
thermal power plants 212
thermal power units 177, 228
thermal renovation of buildings 50, 52, 53, 57
third Conference of Parties (COP3) 35, 37
total cost method 5
transmission grid systems: from generation businesses 217, **218–219**, 219; rules for 223
transmission systems operator (TSO) zones 217, **218–219**, 220, *220*, *221*, **222**, 224, 230
Treaty on the Functioning of the European Union (TFEU) 114
trend-setting model of local development 245
TSO zones *see* transmission systems operator (TSO) zones

UC-ELD model *see* Unit Commitment with Economic Load Dispatching (UC-ELD) model
UCM *see* unit commitment model (UCM)
UNFCCC *see* United Nations Framework Convention on Climate Change (UNFCCC)
UNFCCC Conference of Parties 38
UN Habitat 199
Union of Oil Workers of the Mexican Republic 246
unit commitment model (UCM) 9
Unit Commitment with Economic Load Dispatching (UC-ELD) model 212, 225–226, 228
United Nations Framework Convention on Climate Change (UNFCCC) 1, 35, 64
United Nations' Sustainable Development Goals 9, 39, 161, 193, 246, 259
urban energy innovations, cross-cultural experiments 204; French cities' energy

planning 208–209; Japanese smart community 205–206; Lyon smart community 206–208
urban planning: in French smart cities, new governance of 202–203; laws 52
urban smart turn 202
urban strategies in France 200; city making and governance, newcomers of 200–202; limits and asymptote effects of 203–204; new governance of 202–203
Utagawa, M. 9

value added to local economies in Japan 122; analysis of 132–133; case studies of 134–137; community energy sector 122–124; customers 131–132; investment 131; municipal power suppliers (MPS) 124, **125–130**; power sources 131; supply and demand management 132; *see also* Japan
Vanneaux, M. -A. 8
variable renewable energies (VREs) 212, 235, *235*
voluntary initiatives and decentralization, French energy system 145–147; ETCs, creation of 154–156; innovative rules of governance 147–149; TEPCVs, state's mistrust and failure of 152–154; TEPOS 150–152
"voluntary" interventions 57–58
VREs *see* variable renewable energies (VREs)

Wade, J. 3
Waita Community Joint Company 75–76, *76*
Wargon, E. 155
WEM *see* wholesale electricity market (WEM)
western Japan grids 212; balance in high load period *238*, 238–239; balance in low load period 231–233, *231–233*; constraints 226–228; control reserve activations in low load period 236, *236*; curtailment (output reduction) of renewable electricity 223–224; development plan of interconnections **224**, 224–225; expansion and development for renewable businesses 219–221, **222**; formulation 226; Fukushima nuclear accident, nuclear and renewable energies and 214–217,

216, **217**; historical trends in electricity generation 212–214, *213–215*; points for analysis 225; renewable electricity shares in high load period *239*, 239–241, **240**; renewable electricity shares in low load period 236–238, **237**; target capacity and assumptions for 228, **229**, **230**, 230–231; transmission grids and interconnections 223; transmission grid systems from generation businesses 217, **218–219**, 219; UC-ELD model 225–226; utilization of inter-regional transmission 233–236; *see also* Japan wholesale electricity market (WEM) 249–251

Wirth, S. 184
wood-fired boiler 170, *170*, *171*, 175–176
WWF Japan 39

Youhyun Lee 7